It's a rare thing for a new planet to be discovered, in our own solar system at least. It's equally rare to witness the emergence of a planet's archetypal significance. Though it will likely be decades before a strong consensus is established in the astrological community regarding the meaning of Eris, Le Grice has surely laid the foundations for future reflection on its possible archetypal import. Written in a style both accessible and engaging, Le Grice displays an impressive multi-disciplinary reach, drawing with equal competence from the sciences and psychology, from mythology and world history. Though focused on the meaning of Eris, the book also serves as a lucid introduction to the new field of archetypal cosmology. At the same time, given his central hypothesis, the reader is presented with a series of illuminating insights having to do with the evolution of consciousness and this current, critical phase of the Planetary Era.

Sean Kelly, author of *Coming Home: The Birth and Transformation of the Planetary Era.*

DISCOVERING
ERIS

DISCOVERING
ERIS

The symbolism and significance of a
new planetary archetype

KEIRON LE GRICE

Floris Books

This book is also available
as an eBook

First published by Floris Books 2012
© 2012 Keiron Le Grice

Grateful acknowledgment is made to Chad Harris
for creating the chart images in figures 8, 9 and 10

British Library CIP Data available
ISBN 978-086315-867-4
Printed in Great Britain
by Bell & Bain Ltd., Glasgow

In memory of my father, Barry Le Grice

Contents

List of Figures

Acknowledgments

Over the last seven years, it has been my good fortune to participate in the Philosophy, Cosmology, and Consciousness graduate programme at the California Institute of Integral Studies (CIIS) in San Francisco. The community of researchers at CIIS are devoted to exploring ideas at the vanguard of contemporary thought and to articulating new conceptions of the nature of reality that provide alternatives to the one-sided mechanistic materialism of the modern era. In an age of compartmentalized academic disciplines and often over-specialized, fragmented thinking, this community provides a rare opportunity to study in unison many subjects that are ordinarily deemed to be largely unconnected – including cosmology and depth psychology, mythology and metaphysics, ecology and religion, art and science – and to relate these to the crises and challenges of our time. This essay has emerged out of many of the ideas encountered here, and from my involvement in the field of archetypal cosmology over the last decade.

I wish to express my thanks to the participants in the CIIS spring 2007 graduate seminar, 'The Planetary Era: Towards a New Wisdom Culture', in which my ideas for this thesis were initially formulated. For helpful feedback on an early version of this work, my thanks to Sean Kelly and Richard Tarnas. This essay has also benefited from stimulating discussions with Joseph Kearns, Grant Maxwell, Frank Poletti, Tom Purton, Jonah Saifer, and Richard Wormstall. My gratitude, too, to Christopher Moore and his colleagues at Floris Books for their help preparing the work for publication.

I also wish to express my thanks to David and Margaret Davies for their generous support, without which it would not have been possible for me to complete this project. Finally, I am especially grateful to my wife, Kathryn, for her loving support and encouragement, and for her meticulous and insightful editing of the manuscript.

KLG, November 2011

The unlike is joined together, and from the differences results the most beautiful harmony, and all things take place by strife.

– HERACLITUS

Preface

The recent discovery of planet-like bodies in the outer reaches of our solar system, beyond the orbit of Pluto, has marked an unusually important period in the contemporary history of astronomy. For the first time in decades, in the wake of these discoveries, astronomers were called upon to revise their view of the planetary makeup of the solar system and to settle upon a new consensus as to the exact definition of what a planet actually is. In October 2006, a new map of the solar system, now comprising eight planets and three *dwarf planets*, was unveiled to the world.[1] In June 2008, this classification was revised again to include the new category of planetary body called a *plutoid*. As it stands today, the solar system is comprised of eight planets and five dwarf planets: Ceres, Pluto, Haumea, Makemake, and Eris (see Figure 1). The last four of these objects are also plutoids, a term reserved for dwarf planets lying beyond the orbit of Neptune (see Figure 2).[2]

Figure 1. The New Solar System: Planets and Dwarf Planets.

Name	Year of Discovery	Category
Ceres	1801	Dwarf Planet
Pluto	1930	Dwarf Planet and Plutoid
Quaoar	2002	Trans-Neptunian object
Haumea	2003	Dwarf Planet and Plutoid
Makemake	2005 (March)	Dwarf Planet and Plutoid
Eris	2005 (January)	Dwarf Planet and Plutoid
Sedna	2004	Trans-Neptunian object

Figure 2. Dwarf Planets, Plutoids and Trans-Neptunian Objects.

Of all these new planetary bodies, the discovery of the dwarf planet and plutoid Eris (originally designated 2003 UB313), some 14.5 billion kilometres from the Earth (three times as far as Pluto and similar in size) was unquestionably the most notable development. Significant in itself, the discovery caused an ever greater stir than it might have otherwise, making headline news in 2006, as it led to the controversial demotion of Pluto from its prior status as a planet to that of dwarf planet.[3]

All this, of course, is widely known. What is less well known, though, is the possible deeper, symbolic relevance of these events. What most onlookers would have been unaware of is the fact that the discovery of Eris might also have an archetypal meaning, which emerges not from the science of astronomy, but from the ancient mythic perspective of astrology. As we will see, this archetypal meaning appears to be suggested, in part, by the symbolism, actions, and character of the classical Greek goddess of strife, Eris (pronounced *ee-ris*), after which the dwarf planet was named.

While many people would be quick to reject outright the truth claims of astrology, for those who care to delve into the deep mysteries of the human psyche the astrological perspective can provide a unique source of insight into the archetypal dynamics underlying our conscious experience. Recent evidence of striking correlations between planetary cycles and the themes and patterns of world history has given to astrology a new, unexpected credibility and provided the most compelling evidence yet that this ancient symbolic system, following decades of reformulation through its encounter with depth, humanistic, and transpersonal psychology, is once again worthy of serious consideration.[4]

Archetypal astrology, as this new approach has been called, is based on the supposition of a correspondence between the planetary order of the solar system and the dimension of the human psyche that C.G. Jung called the archetypes of the collective unconscious – creative ordering factors in the depths of the unconscious that give a dynamic thematic structure to human experience. In archetypal astrology, the Sun, the Moon, and each of the planetary bodies, are thought to represent distinct archetypal principles. Thus the planet Mars, for example, is considered to be related to the archetype of the warrior and more generally to the principle of self-assertion and aggression, whereas Venus, understood in its simplest terms, is related to one aspect of what Jung called the *anima*, and to the universal principles of love, beauty, and pleasure. Rather like the ancient mythic conception of the gods, and as in the Platonic conception of archetypal Forms, the archetypal factors studied in astrology are recognized to be not only psychological, but also cosmological in essence, exerting a dynamic formative ordering influence on both the interior and exterior dimensions of reality.[5]

As I have explained elsewhere, the archetypal approach to astrology must be distinguished from the fatalistic predestination long associated with astrology's traditional and popular forms. According to Richard Tarnas' helpful definition, astrology is not *literally predictive* of actual future events and therefore indicative of the inescapable workings of a preordained fate; rather, it is *archetypally predictive* in that its methods of analysis and interpretation of the planetary positions and movements give insight into the archetypal determinants, the general themes and motifs, evident in our experiences, rather than the specific form of manifestation of these archetypes. To understand how an archetypal factor might manifest in the particular details of human life, one would need to take into consideration many other factors not apparent from astrology alone, such as cultural background, gender, economic and social conditions of the time, genetic inheritance, and, most especially, the degree of conscious awareness informing our actions and decisions. The new archetypal understanding of astrology is informed by a fundamental insight into the complex participatory nature of human experience. It is based upon the recognition that the expression of archetypes is shaped and transformed by human consciousness. Archetypal principles, as Tarnas points out, are both *multivalent* and *multidimensional*, which means they manifest in a wide variety of ways while remaining consistent with their core meanings, and they take different forms of expression across all dimensions of reality.[6]

I should explain also that while astrology is incompatible with the basic tenets of mechanistic science and the materialistic conception of the nature of reality that have prevailed throughout the modern era, it is far more congruent with many of the *new paradigm* perspectives that have recently emerged in physics, biology, psychology, and elsewhere. The ideas of holism, interconnectedness, interdependence, organicism, self-organization, and non-local causality that have emerged from relativity theory and quantum theory in physics or from the systems approach in biology have presented us with a view of reality sharply divergent from that based on classical physics and the still-dominant Cartesian-Newtonian mechanistic paradigm. As I have attempted to show in *The Archetypal Cosmos* (2010), these new models, together with the insights of the psychology of the unconscious, provide an increasingly coherent and supportive theoretical context within which we can better comprehend the likely basis of astrological correspondences. It is in this context that I offer the following reflections on the possible archetypal and evolutionary meaning of the discovery of Eris.[7]

Introduction

Within the astrological community, it is now widely believed that, besides its relevance for astronomy, the discovery of a new planetary body in our solar system also has a psychological or even spiritual significance in that it appears to herald the decisive emergence of an archetypal principle into the foreground of our collective experience – a principle previously unrecognized in the astrological pantheon. If so, the discovery of a new planet entails not only the revision of the map of the solar system in astronomy, but also the revision of the map of the archetypal composition of the human psyche in astrology. For both astronomers and astrologers alike, then, the discovery and subsequent naming of Eris was a development of considerable interest, immediately stimulating conjecture and debate among those conversant with astrology as to what the archetypal meaning of Eris might be.

For most of us who were not around to witness the last major event of this kind – the discovery of Pluto in 1930 – the discovery of Eris affords us a unique opportunity to observe and examine, and perhaps even contribute to, the process by which the archetypal meaning of a planetary body is established.[1] Potentially, we are now in the fortunate position of being able to witness firsthand the emergence of an archetypal principle into the collective consciousness and thus to contemplate its possible psychological, spiritual, and evolutionary significance for human life.

The symbolic attitude

Of course, even to entertain the possibility that the discovery of a new planetary body is in some way symbolically meaningful demands a radical deviation from our usual mode of interpreting reality.

Ordinarily we are concerned only with the literal or factual value of things and events. From this perspective, planets, asteroids, and moons are just great masses of rock, ice, or gas that do not signify anything beyond their concrete, physical being. For the modern scientifically educated mind, trained to attend to hard objective facts and to dispel all traces of projected mythic, symbolic thinking, the idea that there might be some unsuspected psychological significance associated with the planets seems quite absurd. To appreciate the arguments to follow, therefore, I must call upon the reader to take a different interpretive posture – to assume, as a complement to rational analysis, what Jung called the 'symbolic attitude', in which one adopts 'a definite view of the world which assigns meaning to events, whether great or small, and attaches to this meaning a greater value than to bare facts'.[2] Any deeper significance of the discovery of the dwarf planet Eris will only become apparent if, by adopting the symbolic attitude, we remain open to the possibility that this way of interpreting our experience has potential validity and that it might disclose to us something of value, something which sense perception, scientific empiricism, and rational analysis by themselves could not. Maintaining critical rigour is, of course, essential to this endeavour: To perceptively distinguish genuine archetypal meaning from our emotional projections and delusions, and to guard against the universal human tendency to see what we want to see, acute discrimination and honest self-appraisal are indispensable virtues. For indeed, archetypal meaning is not (or at least it should not be) merely something constructed in the individual human imagination; it is not the result of a subjective projection of meaning by an isolated interior psyche onto the external cosmos, itself essentially meaningless. Rather, it can only be properly understood, I believe, as the recognition and discernment of what seems to be some form of *objective meaning* inherent in nature, the revelation, perhaps, of the interior meaning of the cosmos itself – of what the ancients called *anima mundi* or world soul.

Six approaches

As we have seen, in astrology each of the planetary archetypes is associated with a range of related meanings, themes, impulses, and drives. In considering the nature of the archetype associated with Eris, then, we are looking to identify another such principle, distinct from the others in the existing astrological pantheon, and the set of themes and a set of characteristics associated with this principle.

With this in mind, let us begin our exploration of the possible significance of Eris by identifying first the main approaches through which the archetypal character of any newly discovered planetary body might plausibly be determined. By considering the archetypal meaning of the existing planets, and by looking back to the discoveries of the three other modern planets (Uranus, Neptune, and Pluto) to see how astrologers retrospectively determined the meaning of these, we will then be able to explore how we might now establish the themes and qualities associated with Eris. For those involved in the fields of archetypal cosmology and astrology this might, I hope, provide a useful starting point for further research in this area. More generally, if we can gain some insight into the possible archetypal significance of Eris this might also enable us to cast a light on the meaning of our moment in history at this most critical stage of what the French philosopher Edgar Morin has called the Planetary Era.[3] If we can understand the archetypal principle associated with Eris, perhaps we will be able to better comprehend the deeper meaning of the early twenty-first century and gain some insight into how best to meet the unprecedented challenges we now face.

As I see it, there are at least six closely related approaches we might pursue to attempt to ascertain the meaning of a new planet:

1. The archetypal meaning associated with a planet seems to be symbolically suggested by the planet's position in the solar system relative to the Earth, the Sun, the Moon, and the other planets. Thus, we would expect to glean some information about the nature of the archetypal principle associated with Eris by considering the placement of this dwarf planet in the outer reaches of the solar system.

2. There seems to be a synchronistic correspondence between the discovery of a planet and themes evident in the major historical events and the *Zeitgeist* of that time, such that these themes appear to be suggestive of the planet's archetypal meaning. If so, then a consideration of the major events of

recent years might throw some light on what types of themes Eris might be associated with.

3. Related to this, the discovery of a new planet appears to indicate a greater recognition and differentiation of the archetypal principle associated with that planet in the collective psyche; and this in turn seems to be significant for the evolution of consciousness and culture – for understanding the course of our collective psychospiritual development. We might thus consider what significance, if any, the discovery of Eris at this particular time has for understanding the current evolutionary challenges we now face.

4. An examination of the other planets suggests that there is a connection between the archetypal meaning of a planet and the mythological associations with the planet's name, which might be disclosed through an exploration of myths featuring the corresponding god or goddess, through intuitive sources such as revelation, dreams, or active imagination, as well as through works of art and literature. By considering the myths featuring the goddess Eris, therefore, we will consider how these myths might illuminate the meaning of the Eris archetype in astrology.

5. It is evident from our knowledge of Uranus, Neptune, and Pluto, that an understanding of the themes and qualities associated with a planetary body can be augmented by considering relevant sources from philosophy, religion, and science, as well as considering how this principle expresses itself in history. Indeed, what will prove to be decisive in establishing the meaning of Eris is the empirical research by astrologers into the role of this planetary archetype in history and culture.

6. Within astrology itself, we might also consider how the new dwarf planet and its associated archetypal themes might be incorporated into existing astrological theory. The discoveries of Uranus, Neptune, and Pluto

necessitated a revision of the theory of planetary rulership by which each planet is associated with a particular sign of the Zodiac and its associated 'house'. We might thus also explore how this system would need to be further revised to accommodate Eris and the other newly discovered planetary bodies.

Let us now consider each of these approaches in more detail.

1. The Symbolic Meaning of a Planet's Position in the Solar System

Whether they are ancient or modern, 'the basic meanings of the planets', as astrologer Dane Rudhyar observes:

> belong to them by the *logic of their positions in relation to the Earth and by virtue of their astronomical characteristics*, such as speed of revolution and rotation, mass, color, number of satellites, etc.[1]

While it might be too much to say that the astrological meaning of the planets is actually determined by their physical positions and attributes, there seems to be a deep symbolic connection between the solar system and the collective unconscious psyche (which will become more apparent as we proceed), such that the physical characteristics and position of a planet within the solar system are in some way suggestive of the meaning of the corresponding *planetary archetype* recognized in astrology. These planetary archetypes can be conceived in Jungian terms as 'formative principle[s] of instinctual power' that give a thematic predisposition and organization to human experience.[2] They seem to relate most especially to Jung's later conception of the archetype *per se* and also to his notion of the *psychoid* nature of the archetypes, by which he sought to convey something of their complex essence as principles that are at once both cosmological and psychological, manifest in the materiality of the cosmos yet giving rise to archetypal images and mythic motifs in the psyche.[3] The planetary archetypes recognized in astrology appear to be universal principles lying behind the more specific *archetypal images* identified by Jung – such as the anima, the shadow, the hero, and the wise old man. The astrological Saturn, for example, includes within its more general, universal meaning aspects of the Jungian archetypes of the wise old man, the father, and death, as well as the Freudian concept of the superego and the principle of time. And the Moon in astrology is related to the three Jungian archetypes of the anima, mother, and child. These archetypal images and concepts, which are overlapping and mutually implicated, are best understood as derivative expressions of the underlying planetary archetypes. The gods and goddesses of mythology, similarly, appear to be expressions of the core archetypes *per se*, personifications and inflections of the deeper universal principles associated with the planets.

The luminaries and the inner planets

The symbolic correspondence between a planet's physical characteristics and its archetypal meaning is most clearly evident with those planets closest to us here on Earth and, most especially, with the Sun and the Moon. Just as the Sun is the burning fireball which gives off the light and heat that makes life possible on Earth, so in the world's symbolic traditions the Sun's archetypal meaning is associated with the creative life force, with the light of consciousness, with life energy and vitality, with the urge to be and to shine. Reflecting its position as the central star of the solar system, the Sun symbolizes the centralizing principle of the psyche, the principle of identity and luminous conscious selfhood. Rising each day out of the ocean of unconscious darkness, the Sun is also associated in myth with the challenge of the hero to bring light, consciousness, and a new flow of life to the world.

In complementary relationship to the Sun, the Moon, as the dominant nocturnal light, is associated with the light within the darkness of the unconscious psyche itself, and with the encompassing, generative matrix of being out of which solar consciousness is born. With its twenty-nine day orbit around the Earth, it has a symbolic correlation with the menstrual cycle and thus, by inference, with women, motherhood, and nurturing. Just as the Moon's light is dependent upon and reflects the light of the Sun, so the archetypal meaning of the Moon includes all types of dependent relationships and anything to do with reflection or response such as one's emotional reactions, one's feelings about life, and one's self-image.

Mercury's archetypal meaning, too, is reflected astronomically: the planet's close proximity to the Sun, its small size, and in its swift movement and speed of orbit around the Sun all bring to mind the character of the Roman wing-footed messenger god after which it was named (the Greek Hermes) and, indeed, the astrological principle is itself associated with communication and movement, as well as perception, knowledge, information, and understanding.

Because of their greater distance from Earth and their small size in the night sky, it is far more difficult, if not impossible, to perform this kind of symbolic analysis on the other planets. What does seem to be applicable to every planet, though, is a symbolic reading of its *position* within the solar system. As I argue in *The Archetypal Cosmos*, there seems to be a symbolic correlation between the *expanse* of outer space and the *depth* of the psyche (metaphorically suggested by terms like 'deep space' and 'the depths of space') such that the distance of the planets from the Sun appears to reflect the 'depth' of these archetypal principles in the psyche and, consequently, indicates how difficult it is to bring them into relationship with the consciousness and to constructively integrate them into human life.[4] In other words, the physical distance of the planets from the Sun seems to relate to the psychological 'distance' of the corresponding archetypal principles from consciousness.

Accordingly, astrologers have recognized that the luminaries (the Sun and the Moon), and the inner planets (Mercury, Venus, and Mars), positioned relatively close to us here on Earth, are related to the basic dynamics of the human personality, such as conscious identity and selfhood (the Sun), the emotional, feeling-based dimension of the personality (the Moon), intelligence, communication, and perception (Mercury), pleasure, romance, evaluations, and aesthetic responses (Venus), and self-assertion, aggression, and the impulse to action (Mars). Although these archetypal principles possess a relative autonomy and are never entirely under our conscious control (we do not, for example, call up anger at will or choose whether or not we like something), these principles are, nevertheless, immediately accessible to consciousness and integral to the functioning of the personality across all dimensions of human experience.

Jupiter and Saturn

Moving outwards, away from the Sun and past the asteroid belt beyond Mars, the planets Jupiter and Saturn symbolize archetypal principles that, astrologers have suggested, form a 'bridge' between the personal and the collective realms of experience. Again, in the context of human experience these principles relate to the individual's relationship to society and the wider culture. Jupiter represents the archetypal principle of expansion, of elevation, of breadth of vision, and it therefore refers, when applied to the individual life, to the impetus to move out beyond the sphere of one's personal concern by expanding the scope of one's awareness and interest to incorporate more and more of the wider world. Saturn, in many respects the opposite in archetypal meaning to Jupiter, represents the principle of contraction, boundary, and limitation. It relates to the structure of the world, to the established order, to one's participation in society through the assumption of the responsibility or burden of some definite role, and the acceptance of the limitation and pressure this inevitably brings. In more explicitly psychological language, Saturn is understood to refer as well to what Alan Watts famously called the 'skin-encapsulated ego' – to the limits of the individual personality, to one's ego-boundaries that arise in connection one's worldly identity and the material pressures of individual selfhood. With Jupiter and Saturn, then, we can observe a transition from those archetypal principles relating to the basic dynamics of the human personality to deeper principles that pertain to the individual's place within the fabric of society, civilization, and the world order, as well as to the corresponding dimensions of the structure of the human psyche.

Uranus, Neptune, and Pluto

Even deeper into the solar system, lying beyond the orbit of Saturn, the distant outer planets Uranus, Neptune, and Pluto represent archetypal principles of tremendous potency and transformative potential whose nature is such that it requires acute self-knowledge and prolonged effort to bring these principles into a constructive relationship with consciousness and to come to terms with the profound psychological transformation elicited by these forces. Far removed from the purely personal sphere of human life, these principles nevertheless exert a far-reaching and often dramatic influence on us from the depths of the unconscious psyche. In the case of Uranus, we find that just as the *planet* Uranus is the first planet beyond Saturn in our solar system, the first planet that is more distant from us than Saturn, so the Uranus *archetype* is associated with the experience of going beyond the traditional Saturnian threshold: it relates to the rebel or revolutionary who upsets the established order; it is creative intuition and invention that breaks through existing limits of knowledge; it is the Promethean energy that pushes humanity towards liberation and freedom from all forms of limitation and restriction; it is associated with the power of the unconscious psyche to effect radical unexpected change in society, to bring a sudden reversal of the established order, and to disrupt the orderly world of ego-consciousness. Uranus symbolizes the restless, highly charged energy breaking out of the unconscious that cannot easily be contained in the conscious structure of the personality, and that promotes creative adaptation, or perhaps brings a troublesome, disruptive neurosis. In all these senses and more, the Uranus archetype leads humanity beyond the established limits of normal experience associated with the Saturnian dimension of experience. When encountering the Uranian energy, one is subject to a principle that is decidedly not of one's own volition, that originates from beyond the personal or social-collective realms of human experience, that is bestowed as a creative gift (as in the 'eureka moment' of scientific discovery) or is thrust upon one even against one's conscious intentions (as in a blunder, accident, or chance encounter that unexpectedly alters one's life direction).

Planet and Symbol	Archetypal Meanings and Associations
The Sun ☉	The light of consciousness; selfhood and identity; the central principle of will and intention; conscious awareness; life energy and creative power; individuality and the urge to self-expression; the hero, the animus, and yang qualities.
The Moon ☽	The feelings, emotional responses and reactions; the inner self; the principle of relatedness and care, nurturing and dependency; the mother and the child; the family, the home, and the past; the womb, the matrix of being, the anima, the Great Mother, and yin qualities.
Mercury ☿	Thinking, perception, communication, and knowledge; the urge to understand and explain; the intellect, analysis, and information.
Venus ♀	Romantic love, beauty, pleasure, attraction, harmony, the urge to please and be pleased, and to give and receive affection, the aesthetic sense, the anima (in its connection with the inner image of the ideal feminine).
Mars ♂	Self-assertion, physical energy, action, fight and struggle, courage and initiative, strength and aggression, the archetypal warrior and the animus (in its connection with the inner image of the ideal masculine).
Jupiter ♃	Expansion, magnitude, amplification, elevation, growth, bounty and abundance, optimism and trust, the urge to self-improvement, the impulse for greater breadth of experience, the desire to connect to larger wholes.

Planet and Symbol	Archetypal Meanings and Associations
Saturn ♄	Contraction and restriction, structures and boundaries, the traditional and the established, pressure and discipline, limitation and fear, duty and responsibility, authority and judgment, old age, time and mortality, death and endings, crystallization of form.
Uranus ♅	Freedom and individualism, rebellion and revolt, liberation and emancipation, sudden unexpected change and reversal, the new and revolutionary, invention and creative genius, awakening and the experience of breaking through.
Neptune ♆	Transcendence and spiritual experience, dissolution and synthesis, oceanic oneness and undifferentiated unity, the ideal and the imaginary, myth and dreams, the enchanted and the sacred, the elusive and the illusory, the subtle and the sensitive.
Pluto ♇	The primal power of destruction and creation; empowerment and intensity; unconscious compulsion; evolution and transformation; instinctual energy; the 'underworld' of the repressed unconscious; the natural cycle of birth, sex, death, and rebirth.

Figure 3. Table of Planetary Symbols and Meanings.

Because it is the most distant planet visible with the naked eye, for centuries Saturn was considered to be the outermost planet, marking the edge of the solar system. The planets beyond Saturn only became known with the aid of the telescope, further suggesting their association with archetypal principles that are deeper, concealed, and often more unconscious in their forms of expression. As with the outer planets themselves, the corresponding planetary archetypes lie outside our ordinary view; these archetypal principles represent forms of transformation that are concealed from our normal conscious awareness. We can see here again the symbolic association of Saturn with boundaries, with the threshold of ordinary human experience. Indeed, in myth, the Saturn principle is often symbolically represented as the 'guardian of the threshold', the fearsome presence that marks the gateway to the *transpersonal* domain of the psyche. In this context, the Uranus archetypal principle (together with the archetypes of Neptune and Pluto) is that which impels us to go beyond Saturn, as it were. Uranus is a transpersonal impulse whose ultimate purpose seems to be to liberate humanity from the old, the rigidly crystallized, and the life-resistant established forms that obstruct creative change.

Beyond the orbit of Uranus, moving farther towards the outer reaches of our solar system, we arrive next at Neptune and Pluto. Whereas Uranus represents the principle that sporadically breaks through the Saturnian structures and limitations, the archetypal Neptune acts to completely dissolve all defined structures and separate boundaries, leading towards the undifferentiated state of 'oceanic' oneness or to the rapture of transcendent illumination of spiritual-mystical experience. It refers to the world of the mythic and archetypal imagination, the realm of dreams and fantasy, that underlies the Saturnian order, and that shapes and ultimately encompasses entire civilizations through their foundational myths and religions. Pluto, which is usually even more remote than Neptune, represents what is, in certain respects, a deeper principle still – the primal power of destruction and creation on which the cycle of all life experience depends, and the evolutionary dynamic within nature that continually brings about the radical metamorphosis of all forms. The archetypal Pluto represents the power that can totally destroy and annihilate all existing life structures, as in the catastrophic eruption of a volcano, or a similarly catastrophic eruption of the unconscious psyche, vividly demonstrated by the twentieth century's world wars.

The outer planets, because of their greater distance from the Sun, obviously have far longer durations of orbit than the inner planets. Consequently, whereas the luminaries and the inner planets move quite quickly through the Zodiac (the primary frame of reference used in astrology) completing one full cycle in anything from one month to two years, the outer planets take decades to fulfil their orbits. Uranus, for example, takes 84 years to complete its cyclical journey, and Pluto takes 248 years. Accordingly, the activated influence of the corresponding archetypal principles associated with the outer planets is of a longer duration, instigating more enduring, more radical, and more profoundly consequential changes in human experience than the archetypes associated with the quicker moving inner planets. The outer planetary archetypes are often unconscious and impersonal in their modes of expression, and are more closely connected with the unconscious ground of human experience than the other planetary archetypes.[5] Although all the planetary archetypes are intimately associated with and ultimately rooted in the collective unconscious psyche, the principles associated with the three outer planets seem to be particularly connected with the essential dynamics of the psychological transformation process that Jung called *individuation*. The archetypal principles associated with the outer planets are, at once, powers of immense evolutionary, creative potential and of immense, life annihilating destruction. They are instinctual and archaic in essence, and yet also transformative, spiritual, and progressive. They can act to sensitize, inspire, and deepen our conscious experience, but also to obliterate, dissolve, and disrupt. The transpersonal archetypal principles relate to the transforming power of the ground of being; and whether this power is experienced as a force for good or ill seems to depend, at least in part, on the conscious attitude of the individual.

Planet	Duration of orbit around Sun
Earth	365 days
Moon	28 days
Mercury	88 days
Venus	227.4 days
Mars	687 days
Ceres	4.6 years
Jupiter	11.86 years
Saturn	29.45 years
Chiron	50.76 years
Uranus	84.02 years
Neptune	164.79 years
Pluto	248.09 years
Haumea	281.93 years

Quaoar	285.97 years
Makemake	305.34 years
Eris	557 years
Sedna	10500-11809 years (estimates)

Figure 4. Duration of Orbits of the Planets.

The position of Eris

Reflecting on the meanings of the planetary archetypes in relation to the sequence of the planets in the solar system, then, we can clearly discern an inherent logic to the planetary order. With increasing distance comes increasing depth: The more remote a planet is from the Sun, the deeper and more consequential are the transformative effects of the corresponding planetary archetype and the more likely it is, therefore, that this principle will remain as an unconscious determinant behind human experience.

If this logic holds for every planetary body within our solar system, as seems likely, we would expect the new dwarf planet Eris, whose orbit reaches beyond the region known as the *Kuiper Belt*, into the so-called *Scattered Disc* area of the solar system, and thus far beyond the orbits of Neptune and Pluto, to symbolize a transpersonal archetypal principle that is even deeper or more fundamental in its effects and its meaning than Pluto.[6] Given its extremely long duration of orbit (some 557 years as compared to Pluto's 248-year orbit), it seems possible that Eris might be associated with a mode of transformation even more profound, more consequential and yet perhaps more archaic, primitive, and deep-rooted than those associated with the other outer planets. When considering the meaning of Eris, then, we are probably looking for a principle that relates to the deepest strata of the unconscious psyche, a principle, perhaps, that is the basis of the other archetypal principles discussed above, even underlying the evolutionary drive and creative-destructive force associated with Pluto. In the chapters to follow, we will attempt to understand just how such a principle might be expressed, and how it might relate to the other planetary archetypes.

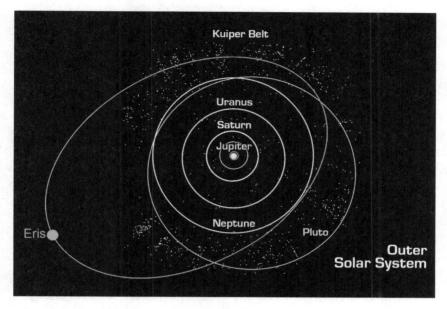

Figure 5. The Orbit of Eris in Relation to the Outer Solar System.

2. Synchronistic Correspondences with Historical Events

Remarkably, the symbolic correspondence between the cosmos and the human psyche is evident not only in connection with space (that is, with the structural organization of the solar system, as discussed above) but also in relation to time, in that the character of the historical period in which each of the modern planets was discovered seems to bear a striking correspondence with the astrological meaning of the discovered planet. From the subsequent study of their meaning in astrological charts, it is now well known to astrologers that the first observation of each of the modern planets (Uranus, Neptune, and Pluto) coincided with a period in history where the themes and characteristics associated with the corresponding archetypal principle were evident both in the prevailing *Zeitgeist* and in certain of the major historical events of that time.[1] The paradigmatic example is that of Uranus (which, as we have seen, is associated with revolution, the overthrow of the established order, liberation, freedom, technological invention, progressive change, humanitarian ideals, and individualism), and the synchronicity between the discovery in 1781 of this, the first of the modern planets, with the American Revolution, the French Revolution, and the beginning of the Industrial Age – each of which, it later became apparent, reflected something of the essential nature of the Uranus archetype.

Synchronicity and planetary discoveries

According to Jung's most precise definition of the term, synchronicity is the 'meaningful coincidence' of an external event and an interior, subjective experience, occurring simultaneously, in which the external event is clearly related to the individual's psychological state at that moment.[2] Synchronicity is the unexpected, uncanny, and often numinous collision of the inner and outer worlds at a precise moment in time for which there seems to be no linear causal explanation – a phenomenon that appears to point to a transcendent or objective dimension of meaning inherent in the nature of things. Much has been written about synchronicity both by Jung and later commentators, and it would be superfluous to repeat that here. A few words are in order, though, about the application of this concept to our theme – the 'coincidence' of the discovery of a new planet (the external event) and the archetypal meanings evident in the culture at that time (the subjective, psychological parallel) – as this presents us with unique challenges requiring a certain adjustment in the application of Jung's definition, on several counts.

First, when applied to collective human experience, and to the entire sweep of human history rather than individual biography, a more expansive understanding of temporal correspondence is required. Viewed against the backdrop of cosmic evolution, a year or even several decades are but moments in time. When considering the synchronicity with a planet's discovery, then, the requirement of 'simultaneity in time' will be applied over this broader period of decades. This broader period is consistent with the observed correlations between historical phenomena and the discovery of the other outer planets in the modern era. A second important difference is that this type of synchronicity is based on our *collective* experience and the *collective* psyche, in which the perception of subjective meaning is to be recognized by many members of the global community and not just one individual person as in the classic cases of synchronicity. A third, further complication is that the meaning of an event, or historical era, might only become truly apparent long after the time it took place. We might be so immersed in the *Zeitgeist* that we are unable so see it in any kind of perspective. Any meaning attached to the discovery of the new planet might remain unconscious to those living at the time until it can be seen in retrospect by future generations. Thus, although the discovery of a new planet is entirely unexpected, as is usual in cases of synchronicity, there is no *instant* recognition of corresponding meaning. Rather, meaning has to be discovered and consciously articulated; and it is to this challenge that we will be giving our attention.

Of all the diverse historical events occurring in any time period, what seems to be significant in observing this type of synchronistic correlation between planetary discoveries and history are only those events that set the era apart from other earlier time periods and not necessarily the most immediately conspicuous events. For example, occurring within two years of the discovery of Neptune, the 1848 revolutions across Europe were, without question, among the most significant events of the time. If one had no prior knowledge of Uranus' association with revolutions, one might well be led to the mistaken conclusion that it is in fact the archetypal Neptune and not Uranus that is associated with revolutions, the overthrow of the established order, and so forth. It is not enough, therefore, that an event is historically significant; it must also be in some way markedly different and unique to that time, in a way that the revolutions of 1848, following on from the wave of revolutions in the 1780s, were not.[3]

We can see this clearly in the case of the discovery of Pluto in 1930 – coinciding with the first harnessing of nuclear power (the first nucleus of an atom was split in 1932), with the marked prevalence of fascism and dictatorships at that time (Stalin, Mao, Hitler, Mussolini), with the collective psychic 'possession' behind the eruption of mass violence during the century's world wars, with the era best-known for Mafia and underworld syndicates in connection with the Prohibition of 1920–1933, and with the exploration of the darker, buried aspects of the human unconscious (Freud's *Civilization and its Discontents* was published in 1930 as his ideas became more widely disseminated through the culture) – events which, as we can see, all occurred in the years and decades close to the time of the planet's discovery. Such occurrences were not only among the most significant of that time in history, they were also in some sense *newly emergent* into the collective consciousness, and relatively distinct from the events of the immediately preceding era.

We see this with Neptune, too, as the discovery of the planet, in 1846, coincided exactly with the 'discovery' of the realm of unconscious psyche by C.G. Carus in the same year (this was the first reference to the unconscious in a textbook on psychology), an event with no prior equivalent in human history (although there were, of course, clear antecedents) and that is directly connected to Neptune's association with the transpersonal dimension of reality and the collective unconscious. Also accompanying Neptune's discovery, and falling under the thematic scope of the Neptunian archetype, as Richard Tarnas notes, were unprecedented developments in photography and moving pictures, developments that ultimately provided the foundation for television and cinema, giving people the power to create, transmit, and manipulate images such as they had never had before.[4] Other developments Tarnas cites as relevant here are the ascendancy of socio-political visions of utopia (*The Communist Manifesto* by Marx and Engels was published in 1848), the beginning of the pharmaceutical industry, the first use of anaesthetic, the spread of Spiritualism, the rise of Theosophy, the full flowering of late Romanticism, and the first significant influx of Oriental religious ideas to the West. It was at this time as well that the field of mythological studies began to flourish with Adolf Bastian, writing in the 1870s, positing the existence of universal 'elementary ideas' that are present, he proposed, in the mythological motifs of all the world's cultures – a concept that was later to be influential on Jung as he developed his notion of archetypes. These occurrences, more subtle in character than the events coinciding with the discoveries of either Pluto or Uranus, perhaps reflect the subtle, intangible nature of the Neptunian archetypal principle itself. We can also see that the Neptunian association with spirituality, imagery, idealism, our collective unity, and so forth – even allowing for the pervasive materialistic and positivistic ethos of the nineteenth century – is clearly evident in the events around this time.

Possible correlations with the discovery of Eris

We must remain mindful, then, that it is not necessarily the more overt events of a period in history that bear a synchronistic correspondence to the discovery of a new planet. Given this proviso, and given the difficulties involved in determining just which events are to be considered characteristic and particular to a given period of time, it might be too soon to make any informed judgments about our own time. But, even from our limited vantage point, it does seem clear that there are two new themes and classes of events that have recently announced themselves on the world stage and that have forcefully emerged into collective awareness. The first is global terrorism and the 'war on terror', as its corollary, which has led to a perilous deepening of tensions and divisions between the Christianized West and certain radical factions of the Islamic religion. The second is the ecological crisis and likely catastrophe we now face, with the threat of the collapse of the biosphere, a possible mass extinction of many species in the not-too-distant future, and – if certain forecasts prove well-founded – a real risk to the continued existence of the human race. With this, somewhat belatedly, we have witnessed a new, widespread awareness of environmental concerns and the recognition of the need for urgent action to counteract the climate change created, at least in part, by human activity. It seems probable that, looking back on the current period from the future, we would likely identify both these phenomena as the defining events and challenges of our time. In addition to these, there are, of course, many other factors and developments that one might point to as characteristic of our time: the global financial crisis; the technologically-driven 'information age' of instant, mass communication and data-transfer around the globe; the development of genetic biology (the first draft of complete human genome sequence was published in 2003) and the power this will surely give us in the near future to intervene and control the biological characteristics of future generations with the creation of so-called 'designer babies'; the unrelenting and rapid pace of technological advancements that will furnish us with the power to increasingly interfere with and dictate the processes of nature across all areas of life; the alarmingly superficial nature of contemporary Western, and now global, culture with its obsession with image and distracting novelty. Each of these – to give but a few choice examples – are potentially important, and our consideration of the meaning of Eris here and in subsequent chapters will take into account the possible significance of these other factors.[5]

Now, just as astrologers have retrospectively observed correlations between the meanings of the outer planets and the events of the periods in which they were discovered so, conversely, we might in this case analyse the events contemporaneous with the discovery of Eris to determine and abstract the possible meaning of the corresponding planetary archetype. Let us consider, therefore, what, if anything, these events have in common. Is it possible to discern underlying archetypal themes inherent within these different factors?

Globalization and reactions against civilization

One way to begin to conceive of possible common themes connecting the major events of our time is in terms of globalization.[6] Obviously the ecological crisis, terrorism, and the financial crisis are all global phenomena whose causes and consequences are intimately connected with the emergence of an industrially powered global economy and the communications network now spread around the entire planet. However, it seems unlikely to me that the archetypal meaning of Eris relates to the phenomenon of globalization in itself as this seems to be adequately represented by other planetary archetypes. In its association with expansion and the movement towards larger wholes, Jupiter is the planetary archetype most clearly connected to globalization. But Neptune, too, in its association with unity and the dissolution of boundaries, has a clear connection to globalization, as does Uranus, in its association with the technological innovation that has supported the emergence of the global economy and culture.

Yet Eris might still be related to globalization in one other respect. Given the especially troubling nature of these crises, it seems possible that the Eris archetype is in some way associated with certain problematic aspects of our experience relating to globalization that must be consciously faced and overcome as we move towards some form of sustainable planetary civilization and planetary consciousness. It could be, in other words, that the events of recent times reflect an archetypal dynamic that is connected to the challenges to be met, and processes to be worked through, as we move towards the realization of the global unity of the planet.

Recalling here our earlier suggestion, that Eris will perhaps symbolize an archetypal principle rooted deep in the collective unconscious psyche, our attention is naturally directed to the deepest causes of the aforementioned contemporary events. Now, both global terrorism and the ecological crisis seem to have arisen, it is fair to say, as a virulent reaction to the way civilization has developed, particularly in the modern era. On the one hand, global terrorism is a reaction, perpetrated by certain radical factions of the Islamic religion, against the value systems and ideology of the now global West, against perceived inequalities in wealth and the distribution of resources, and against perceived injustices against the Arab nations, particularly Palestine. On the other hand, the ecological crisis is believed to be an adverse reaction on the part of the planet Earth and its biosphere to human activity and behaviour patterns, especially since the Industrial Age in the developed West. We see here, then, two forms of *reaction against modern civilization* – either by a group that feels excluded or violated, in the case of terrorism, or resulting from the violation of the planet Earth itself, in the case of the ecological crisis. Obviously terrorism itself is not new; it has been occurring for decades with the IRA bombing campaigns in the UK, with the Basque separatist movement in Spain, and with terrorist groups in various other parts of the world. However, what is new is the phenomenon of *global* terrorism. So too environmental and ecological disasters have occurred throughout the history of the planet, but what is new in this case is, first, that the ecological crisis has been caused in large part by the population explosion and human activity that gave birth to the global economy, and, second, that it demands a global, coordinated response if the measures to counteract climate change are to be successful. Both crises, then, to draw together the above reflections, are connected to *forms of reaction* either against the process leading towards a global civilization, or, equally, reactions that impel us towards the achievement of global unity. The financial crisis, too, is another reaction, of sorts, to capitalism, and more precisely to the corporate power, unregulated profiteering, and unchecked greed that created the current economic situation. So potent are all these reactions that here we have three sets of events that, especially in combination, seem to threaten the very basis of civilization itself, and that call into question the legitimacy of humanity's mode of being, our ways of inhabiting the Earth, and of relating to each other.

The separation of civilization from nature

Consequently, to understand the deepest causes of these crises, and to discern their possible archetypal roots, we must look not to any specific external causal factor or guiding ideology, but to the foundational preconditions of human experience itself. Since these crises are global in outreach and have thrown into jeopardy the very basis of the world economy, human civilization, and the biosphere itself, we must, I propose, seek to uncover the principle or impulse that gave rise to civilization in the first place, the principle that impelled the human to separate itself from nature and to seek control, to manipulate, and to impose some form of order on the natural world.

One could, of course, posit as most fundamental a number of different motivating impulses behind human existence: the survival instinct, the will to power, the fear of death, the pleasure principle, the desire to avoid pain. Each of these drives plays a significant part in conditioning the human reaction to and mode of participation with the natural environment, with other people, and with other species. A case could be made for any of these principles as the most fundamental, underlying all the others. For Schopenhauer, it was the blind will of nature; for Freud, the sexual instinct and the pleasure principle; for Nietzsche and Adler, it was the will to power; for Ernest Becker, it was the fear of and denial of death. Yet all these impulses presuppose a fundamental basic condition of life: namely, the occurrence of change, of ceaseless change, and with it the existence of pairs of complementary opposites such as pleasure and pain, life and death, power and submission, victory and defeat, action and reaction, peace and war, order and chaos. It is the existence of these opposites, and the movement of life between them, that underlies all forms of motivation. This is the basic condition of life, more fundamental even than the fact of death and the desire to prevent it.

One fundamental form of manifestation of this movement of life energy between opposites is an antagonism or discord between human conscious experience and nature, and this seems to be especially apparent in the current crises. In fact, a discordant antagonism between consciousness and nature, an inherent tension between these, has defined the human existential situation from the very outset. It is because the human is a self-conscious being, acutely aware of its vulnerability and mortality, struggling to survive in the often hostile world of nature, that our ancestors were, from the very start, compelled to use their reason and conscious intelligence to try to prevent suffering and postpone their demise through the control of nature. By unconsciously responding to this discord, there arose in human experience the impulse to control nature, and in turn this led, over the centuries, to the gradual movement towards civilization, and eventually to the global economy of the modern world. For what characterizes the human condition – what distinguishes us from other species of animals – is that our instinct to self-preservation together with the will to power and pleasure (which we seem to share with the animal kingdom) is coupled with both a highly developed rational intelligence and conscious self-awareness. This has enabled us, by exercising the power of reason, to develop brilliant, sophisticated, and elaborate means both to ensure our survival, and to make our lives more pleasant and more comfortable. Science and medicine, computers and cars, sophisticated weaponry or complex welfare systems are all attempts, in one form or another, to meet these aims. Yet even today, beneath the veneer of culture, the primary condition underlying human civilization remains an unconscious response to the inherent discord between our conscious self-awareness and our biology, between self-reflective rationality and instinct, between the human self and nature. This response has created a schism between human existence and the rest of the planet, between human consciousness and our natural instincts. This schism seems to be an inevitable consequence, then, of the preconditions of our existence. It relates to the human response to the ceaseless change that characterizes all existence, and to the movement of life between opposite tendencies.

It is this inherent discord between our dual nature as both self-reflective conscious beings and instinctually driven biological beings that has been the primary motive force behind the unfolding course of history, and that underlies the current global situation. It is only because we *know* that we feel pleasure and pain, because we are self-aware beings – vulnerable, mortal, with an agonising awareness of our suffering – that the competition for, and manipulation of, the resources of nature was able to assume the exaggerated and sophisticated form that we now see in modern civilization, and that has created the current crises. Out of this discord between reason and instinct the human world has come into being. But it is out of this, also, that human beings have ravaged and plundered the Earth and butchered each other on a scale far beyond that possible by instinct alone. Human beings responded to the discord that characterized their existential situation and acted upon it without ever coming to terms with it psychologically. And now, millennia later, it is this discord that appears to be surfacing in an extreme form with the ecological crisis and global terrorism.

Others, of course, will point to different explanations as to the root cause of our current crises; but, taken to their deepest origins, many such explanations themselves seem to go back to the inherent discord in existence. For example, writing in 1995, cosmologist Brian Swimme pointed out that 'all our disasters today are directly related to our having been raised in cultures that ignored the cosmos for an exclusive focus on the human.'[7] But why the exclusive focus on the human in these cultures? Because our innate instinct to self-preservation impressed upon us the necessity of ensuring our survival by building cultures to meet this need. And why was this instinct to self-preservation able to elicit such a response? Because humans are conscious beings who were able to use their rational intelligence to do something about the strife they experienced by harnessing resources and establishing modes of living to protect themselves in what was (and still is in many respects) a dangerous, uncertain natural environment.

Discord and evolution

I suggest, then, that if we had to ascribe a single principle to describe the root cause of the current world situation, and a root principle behind the human condition, we might point to the *discord* or antagonism in the nature of things, so acutely experienced by human beings – a discord that arises from change, and that draws forth human self-reflective consciousness as an adaptive response. Mythically described by the biblical Fall of Man and the expulsion of Adam and Eve from the Garden of Eden, this discord arises in coincidence with the emergence of human self-awareness, when humans (having eaten of the tree of knowledge of good and evil) find themselves wandering unprotected in an unpredictable, dangerous world.[8] In psychospiritual terms, this discord is reflected within the human psyche itself as the tension between rational ego-consciousness and the instinctual unconscious, which is its origin and ground. More generally, the discord seems to express itself through, or arise from, the tension between all opposite polarities in existence, opposites that have been increasingly differentiated and thrust apart over the course of the last few millennia, and that have now become fully apparent in the world situation: masculine and feminine, science and religion, spirit and nature, humanity and nature, patriarchy and matriarchy, ego and unconscious, self and other. Tensions between these polarities are evident wherever one looks.

To stop our analysis here, however, would be to over-simplify the situation. The problematic consequences of the existential discord, which we now see plainly manifest in the ecological crisis and global terrorism, arise not only because of the separation of the human from nature, and reason from instinct, but because human beings, despite pretensions to rational autonomy, remain in the grip of powerful unconscious forces. This is the essence of the Faustian problem, first illuminated by Goethe and taken up by Jung and others, in which our rational intelligence remains controlled by the instincts and drives of the unconscious psyche. Having partially separated from its instinctual foundations, and though it would like to imagine itself as the master and controller of its destiny, reason must open itself to the service of the greater life of which it is a part or else it will be unconsciously manipulated by the life dynamic it seeks to repress or ignore, with potentially devastating consequences. Science and technology in the hands of human beings still fearfully motivated by the drive to self-preservation and by the desire for pleasure and power are dangerous tools. As Jung clearly foresaw, we must reckon with the unconscious psyche if we are not to destroy ourselves through ignorance and through the might of our own science and invention.[9]

In astrological terms, the psychodynamics of this existential situation are well represented in the existing astrological pantheon: the Sun relates to human consciousness and awareness; the Moon is associated with our feelings and, together with Saturn, to our need for security and protection; Saturn also relates to pain, limitation and suffering, to the fear of death and the self-preservation instinct, and to physical structures and the law and order that have made it possible for civilization to develop; Uranus, in this context, relates especially to the unpredictable uncertainty of life; and, most important, the archetype of Pluto represents the repressed life power in the unconscious psyche – the force of instinct and unconscious compulsion, the elemental power of nature that impels the evolutionary process itself. However, with regard to Eris, if it is in some sense a more fundamental principle than Pluto, as I have suggested, then it seems possible that Eris might represent the inherent discord associated with the tension between opposite polarities in existence, a discord that underlies the activity of the Plutonic evolutionary force of nature.

In support of this, we might consider Schopenhauer's analysis of the relationship of the universal will (relating to the archetypal Pluto) to strife or discord in nature:

> So everywhere in nature we see strife, conflict, and the fickleness of victory, and in that we shall recognize more clearly the discord that is essential to the will ...This strife may be seen to pervade the whole of nature; indeed nature, in its turn, exists only through it: 'For, as Empedocles says, if there were no strife in things everything would be one and the same.' (Aristotle, *Metaphysics*, B5) This very strife is, after all, only the revelation of that discord that is essential to the will.[10]

Therefore, if Pluto is the planetary archetype most clearly associated with the universal will itself, or the manifestation of this will in nature, then it seems possible that Eris might refer to the inherent discord that accompanies, or gives rise to, the activity of the universal will. Other factors we have still to consider may contradict or modify this hypothesis, but if our analysis of the deeper significance of the major crises of our time is correct, Eris could refer to an archetypal principle associated with reactions, discord, contraries, antagonisms, and – to anticipate our discussion of the mythic connotations of Eris – to the experience of *strife*. Considering the deepest causes of the two major, defining events of our time, then, I venture to suggest that Eris is related to the inherent discord associated with the long evolutionary journey towards globalization and the realization of planetary unity, and the potentially destructive reaction of that which has been violated en route to this.

3. The Evolutionary Emergence of an Archetypal Principle

We should be clear, as we continue, that the discovery of a new planet marks a differentiation of an archetypal principle from the other previously recognized ones and *not* the coming into being of a new principle. The discovery indicates that the set of themes, qualities, and impulses associated with the planet will thereafter become more prominent in our individual and collective experience, and, moreover, that the newly identified planetary archetype must be consciously engaged with if we are to avoid its more deleterious forms of expression. When considering the meaning of Eris, then, we must keep in mind that we are looking for a principle that already existed prior to the discovery of the dwarf planet, but that was, most likely, unwittingly conflated with the other archetypal principles in the astrological pantheon, or that was only implicit within them. For example, it is beyond question that the archetypal principle Pluto already existed prior to the discovery of the planet Pluto since this principle was clearly personified by various gods and represented by theriomorphic symbols found in many of the world's mythological traditions: Dionysus, Hades, the Devil, Pan, Wotan, the Uroborus, Shiva, Shakti, Mara, Moira, and many others. Needless to say, Plutonic type events and experiences did not begin with the discovery of the planet Pluto; this discovery indicated only that the power represented by these mythic figures was thereafter more fully available to human conscious participation (in that, for example, we now held in our hands the power of global destruction with nuclear missiles, and that henceforth, the id – the repressed instinctual unconscious – could potentially be consciously recognized in both our individual and collective experience). For as the *planet* became known to us externally, perceived through the telescopic exploration of our solar system, so the corresponding *archetypal principle* simultaneously became more consciously known to us, as if that principle had announced itself to the collective psyche, and made its full debut, if you will, on the world stage. Our knowledge of the principle permits its differentiation: To know an archetypal principle as something separate and distinct is to relate to it as a subject to an object, and thus to bring to an end the unconscious identification with the principle, at least *in potentia*.

Looking back on the process by which the three planets discovered in the modern era were incorporated into astrology, it is evident now that the archetype associated with each newly discovered planet was initially conflated with the archetypal meanings of the other planets, especially the planet positioned next to it, closer to the Sun. In the years following the discovery of the planet Neptune, for instance, the archetypal meaning associated with Neptune appears to have been partly confused by certain astrologers with that of Uranus. Speculations on the meaning of Neptune, even into the twentieth century, connected the planet to turmoil, revolution, restlessness, and science – each of which are actually associated with Uranus.[1] Other astrologers attributed to Neptune phenomena that were subsequently associated more clearly with Pluto, such as wickedness, dynamite, murder, and mutilation. From the first groping speculations as to the meaning of the newly discovered planet, it took several decades for a fairly accurate meaning of Neptune to be arrived at and for a consensus to be reached.

Given these precedents, we might expect in this case that the meanings of Eris and Pluto, in particular, could easily be conflated, and that we will need to carefully draw out the distinction between these planetary archetypes as we proceed. This might mean that we must refine the range of meanings associated with Pluto, to define the meaning Pluto more narrowly within the context of the larger archetypal pantheon comprising Eris.

Prior to the discovery of Eris and the other dwarf planets, few astrologers would have considered the existing astrological pantheon up to and including Pluto to be insufficient to adequately incorporate all the main dimensions of human experience; fewer still would have explicitly recognized the need for another distinct archetypal principle, let alone anticipated its imminent emergence.[2] However, from an evolutionary perspective, the discovery of Eris now compels us to look more closely at why this might have happened at this time in history. What is it, if anything, that the unconscious psyche wishes us to know about our time? What symbol or portent has been heralded? If Eris does symbolize an archetypal principle connected with discord, the separation of the opposites, reactions, and the process leading towards wholeness through strife and antagonism, and so forth, what must we do to consciously assimilate this archetype into our experience?

Planetary discoveries and the evolution of consciousness

The discovery of the other modern planets seems to have marked an impending evolutionary shift in consciousness in which human beings thereafter have had to come to terms with the newly emergent archetypal principle. With each planetary discovery, we appear to have entered another distinct phase in our collective experience, a phase with its own defining archetypal and evolutionary challenge. If so, then we might surmise that the discovery of Eris, similarly, must also place before us an evolutionary challenge of its own.

If we examine this idea more closely, it seems that in certain respects the planet (or pair of planets) we consider to be the most distant in our solar system, the planet that marks the edge of the known solar system at a particular time in history, indicates the archetypal principle that, in the broadest terms, is currently defining our collective human experience at that time. For example, from antiquity until the discovery of Uranus in 1781, Saturn was considered to be the most distant planet, and human collective experience up to the late eighteenth century seems to have proceeded in general accordance with the archetypal meaning both of Saturn and its neighbour, Jupiter.

We have already touched upon themes relating to the Saturn archetype: structures of all kinds (the ego-structure, the skeletal structure, buildings, the structure of civilization and of society); boundaries separating one thing from another; authority, the establishment, and the rule of law and order; repression, control, and self-preservation; acute self-consciousness and fear; time and mortality; pain, death, illness, old-age – and the desire to protect ourselves from these inevitables. The basic Saturnian motivation –

at least before it is transformed during the work of individuation – is that of fear, of self-protection. The Saturnian drive is to try to preserve one's life, one's position, one's well-being, one's happiness, one's material security.

In evolutionary terms, we can, I suggest, identify two parallel and mutually reinforcing developments that have occurred during the time Saturn marked the known boundary of our solar system – the rise of civilization and the separation of the human conscious ego from its preconscious unity with nature. Civilization, according to Freud's definition, 'describes the whole sum of achievements and the regulations which distinguish our lives from those of our animal ancestors and which serve two purposes – namely to protect men against nature and to adjust their mutual relations'.[3] The movement towards civilization, towards an increasingly urbanized world, has reflected humanity's conquest and control of nature and our increased separation from the natural order. The human world has been built upon and, to a certain extent, is set against the Earth and nature. Buildings, towns, cities, and the social fabric of institutions, government, courts of law, and so forth, all function in the modern world to keep human beings safe and protected, as far as possible, from the caprice of the natural world – from illness, natural disaster, extremes of heat and cold, from external attack, and from poverty and destitution – but they have also led to a greater distance between human life and nature. Laws, rules, codes of conduct, the establishment of ethical standards, moral imperatives, and such like, have forced upon human beings restrictions that have prevented them from living instinctively, and that have driven them ever further from their natural state. It was, for instance, upon the biblical Ten Commandments that Judaism, and later the entire Christianized West, was founded. Obedience to the moralistic will of Yahweh, laid down in the Commandments and enforced by fear of his wrath and retribution, brought to an end the Israelites' involvement with the nature religions and deities of Babylon. Concomitantly, as the influence of Judaism spread, the human ego became set against the natural instincts. To follow Yahweh's commandments meant going against natural inclinations, such that reason, consciousness, and self-will were set against the instincts, appetites, and natural impulses. In this way, both within and without, the Saturnian law and moral order was imposed. Protective boundaries were erected between man and nature, between reason and instinct. Anything that transgressed the boundaries of conventional morality was rendered taboo. The natural instinctive mode of existence that characterized primal cultures thus gradually gave way to the more civilized, more controlled, but increasingly unnatural way of being that characterizes modern life.

As civilization developed, and the primal condition of *participation mystique* (the preconscious identity of human consciousness and the natural world) gradually came to an end, human existence became increasingly autonomous and separate from nature. As we have seen, this development was also played out psychologically, in the changing relationship between the human conscious ego and the instinctual unconscious. Just as the ego has emerged from, rests upon, and is set against nature within (the unconscious), so in the physical world, human existence, similarly, has emerged from and is set against nature without. The two processes are inseparable.

All these factors and themes, each relating to the Saturn archetype's association with boundaries, separation, structures and so forth, have defined the progression of human civilization right up to modern era. This is not to imply, of course, that the other archetypal principles beyond Saturn (that is, those associated with Uranus, Neptune, and Pluto) were not active and influential before this time, only that these principles were manifesting within the context of the separation of the human ego from the unconscious and the construction of the civilized world (Saturn). They were supporting this end – although largely unconsciously, as the corresponding planets had not yet been discovered. At this time, the human species was (as it still is) in the process of becoming ever more separate and distinct from nature, in a movement away from a state of *participation mystique* with nature and preconscious unity with its instinctual ground, and this separation clearly relates to the Saturnian principle.

The rise of the structure of civilization and of the urbanized world (Saturn) has gone hand in hand with material growth and geographic expansion, under the guiding ideology of progress, improvement, and development, all of which fall under the archetypal range of Jupiter. The Jupiter archetype, as we have seen, is the principle of expansion and elevation, of breadth of perspective and wide cultural vision. It is associated with the faith, trust, and confidence to explore, to grow, to adventure farther afield, to seek to take in more and more of the world, to achieve success, and to enjoy nature's bounty. The Jupiterian phase of expansion culminated, in a sense, with the recognition of the Planetary Era, which first occurred during the aptly named Age of Exploration (or Age of Discovery), beginning in the fifteenth century, that brought geographic expansion and conquest across the whole globe. In terms of our collective human experience, it was a defining moment when human beings first realized that the Earth itself is a planet. From then on, once the expansion across the surface of the planet was established, and there were no new physical horizons, no unexplored lands to be conquered, the human race was poised for the next phase of its evolutionary journey. As the visionary French scientist Pierre Teilhard de Chardin points out, it is the very fact that the Earth is a globe, a sphere of limited circumference, that prevents the unlimited horizontal expansion of the human species and forces upon us instead the evolutionary development of consciousness or *interiority*, as he called it.[4] In archetypal terms, because there are structural limitations (Saturn) on expansion (Jupiter), and because there is a finite limit (Saturn) on the Earth's abundance and material bounty (Jupiter), this called forth, once these limits were approached, a psychological transformation of human nature in order to come to terms with these conditions.

To summarize briefly what is obviously a complex argument, we can see that with the discoveries first of Uranus, and then of Neptune and Pluto, the leading edge of the human project was no longer only the expansion or the construction of civilization – although this has obviously continued and at a rapid pace. Rather, the defining themes at the vanguard of cultural evolution were henceforth industrial-technological transformation, the emergence of complex city lifestyles, new utopian social ideals, the discovery and mapping of the depths of human interiority, the exploration of the quantum depths of matter, the spread of democratic individualism, the liberation and empowerment of the individual human self. These factors, and many others, each pertaining to the archetypal meanings of the outer planets, were now defining the future unfolding and transformation of the world. Due to the limitations on expansion, human beings have been compelled, often reluctantly, to come to terms with their own individual interior nature; they have also been forced to develop new socio-political systems to better organize the growing complexity of relationships between individuals and between nations; and, as never before, they have been forced to rely on technology to make more efficient use of the resources at hand. The Saturnian dimension, pertaining most especially to the rise of civilization – its structure, its laws, its history – and the increasing separation of the human ego-self from nature, had thus been surpassed, subsumed in deeper, more complex archetypal dynamics. From the late eighteenth century onwards, a new distinctive transformation was upon us as the archetypal qualities of the outer planets had started to come to the fore.

Each evolutionary cultural transition builds on its predecessor. Thus, building upon the achievements of the Jupiter-Saturn phase of the development of civilization that supported the rise of the modern self, the discovery of Uranus then heralded an accelerated phase in the emergence of individuality with the era of the individual culture-genius, of world historical individuals in the creative arts. It was the age, as well, of democratic liberation (as in the French Revolution), and of decisive developments that furthered the social, political, and economic emancipation of the modern individual self. The discovery of Neptune, you will recall, corresponded with movements that transcended this individualism, whether for good or ill, emphasizing the unity of individuals in the larger social organism (as in communism) and/or the underlying spiritual unity of existence (as in Spiritualism, Theosophy, Romanticism, and with the discovery of the collective unconscious). Pluto, we noted, marked the encounter with immense accumulated power in nature, and the challenge of managing the instinctual power that had been repressed into the unconscious during the course of the evolution of the human self. And within this sequence, Eris, we have said, might pertain to the challenge of coming to terms with the discord inherent in the nature of things that underlies human civilization and that impelled human cultural and psychological development from the first. Perhaps the planet Eris relates to a principle that can only be truly comprehended by plumbing the depths of the human psyche, to bring to the surface what lies behind the Plutonic will to power and compulsive patterns of consumption that have characterized the human attitude to nature, especially in the modern era. If, with the discovery of Pluto, human beings were called upon to consciously confront, harness, and responsibly manage the immense power of nature, perhaps now, with the discovery of Eris, we are being invited to come to terms with the discord and tension between opposites that gives rise to the accumulated Plutonic power and unconsciously determines its expression in human affairs. With the discovery of Eris, it might be that we are being called upon to address the consequences for nature, for our planet, of the course taken by human cultural and psychological evolution.

I believe that Eris, to summarize our thesis thus far, might relate to the consequences of the separation of human consciousness from nature – to the challenge of dealing with the compensatory reactionary force that has arisen in response to the emergence of human consciousness and culture. The violent reactions we are now experiencing in politics, economics, and most especially at the ecological level, point to the fact that our world, our time, has become so dangerously one-sided, out of balance, and alienated from the natural world. The processes that have given rise to modern civilization, created modern industrial society, the global economy, and our technological culture, have pulled us into a state of potentially catastrophic disharmony with the natural order of things. Perhaps such a development was inevitable, something that could not have been avoided. Perhaps it has been the unfortunate consequence of the process by which the modern self, with its autonomous ego-consciousness, could come into existence. Either way, it seems that we must all now face up to the reality of a world increasingly out of balance. We are today facing momentous counterbalancing responses from the planet, the natural world, and from the depths of the human psyche to the current state of civilization and our consciousness. We need now to urgently restore a condition of dynamic harmony by aspiring to realize our wholeness and unity, both within ourselves and with the planet at large.

Planetary archetypes and individuation

From a psychological perspective, we might gain further insight into this process of evolutionary transition if we consider the characteristic pattern of psychological development occurring during the course of individuation, a process of psychological transformation that has been explored in great detail by Jung and his followers.[5] Both individually and collectively, human development seems to follow

two distinct phases of transformation. In the first movement, the
individual ego-self emerges out of its primordial matrix, out of its
preconscious identity with nature. Becoming increasingly separate
from nature, from the social group, and from the world, the human
ego leaves behind the state of *participation mystique* that characterizes
primal cultures and infancy.[6] The ego emerges, ever more defined,
separate, and autonomous, and in parallel with this, as we have seen,
civilization is constructed, built upon and often set against nature.
This first movement, then, relates to what Schopenhauer calls
the *principium individuationis*. It refers to that process by which,
according to Schopenhauer, the original unity of the Universal Will
undergoes a process of atomization into separate units of individual
consciousness, into a plethora of individual selves each apportioned
their own quota of will. Ego-consciousness becomes established as a
separate, relatively stable and autonomous centre within the psyche –
the locus of self-reflective consciousness, will, and personal identity.

In the second movement, which Jung calls the process of
individuation (which represents a significant extension of
Schopenhauer's similarly named concept), the rational ego, the carrier
of consciousness, 're-encounters' the unconscious ground or matrix
from which it emerged, and is called upon to relinquish its assumed
position as central controlling authority in the psyche, submitting to
the greater authority of the Self – a totality and a central point within
the psyche that includes both conscious and unconscious, light and
dark, masculine and feminine, good and evil, spirit and nature. In
this way, through a long dialectic process between ego and Self, the
original wholeness of the human psyche is restored, now enriched
and illuminated by self-reflective consciousness, now known by a
conscious human subject.

In astrological terms, when considering the first movement, the
outer planets can be seen to represent the archetypal powers of the
unconscious ground or matrix from which the ego has emerged
during the course of human evolution. In this respect, Pluto is well
symbolized in myth by the *uroboros*, the self-devouring serpent, which
seems to represent the primordial state of instinctual life perpetually
and blindly consuming itself; Neptune can be seen to refer to
the undifferentiated state of *participation mystique*, to the primal
experience of enchantment and oneness with nature but devoid of any
distinct self-sense or conscious volition – the paradisiacal Garden of

Eden; and Uranus symbolizes the unbridled freedom of instinctual expression in this condition of animal grace – a condition of absolute freedom but only at the behest of the instincts – which is only possible before the imposition of the laws, rules, and conventions of human society (Saturn). In this context, then, and considering what we have already proposed about the archetypal meaning of Eris, we might surmise that Eris could be conceived as the principle associated with the movement of life energy between opposite poles that provides the precondition for the emergence of self-reflective consciousness. Eris might be the archetypal principle of discord relating to the movement of life energy between opposites that promotes the movement away from the primal unity of the Universal Will towards plurality, the impulse that, at the deepest level, lies behind the *principium individuationis* by which the individual ego comes into being.

Based on this line of conjecture, I am suggesting that the Eris principle, in effect, drew out our conscious subjectivity from the unconscious, that it called forth, through sheer necessity, the development of rationality and the emergence of the autonomous ego that could orient itself to the external world and try to exert some form of control on the environment. Without this inherent discord and polarity, if our human ancestors had been able to live in uninterrupted paradisiacal bliss in their earthly home, the development of human consciousness would never have happened. Here, in its relationship to the underlying impulse that is behind the evolution of consciousness, we perhaps have a clue to the deeper value and purpose of the Eris archetype, if it is indeed to be associated with the discord inherent in existence.

Turning now to the second movement, when individuation begins in earnest, the archetypal principles of Uranus, Neptune, and Pluto describe those archetypal powers that impel individuation: Uranus incites one to go one's own way in life; Neptune represents the opening to the spiritual dimension; and Pluto is the evolutionary power in nature that seeks its own death-rebirth and transformation. In addition, however, from a depth psychological perspective, these archetypes also describe dimensions of experience that are to be encountered *during* individuation – they are stations on the way of individuation, as it were. To individuate, as Jung has emphasized, one must first adapt to the ordinary demands of living, to the Saturnian dimension of experience: develop a solid ego, find a place in the world, forge a worldly identity grounded in the practical concerns of living. Before individuation can properly begin, one must have sufficiently assimilated the established knowledge and learning of one's culture, and have involved oneself in the life patterns and possibilities offered by one's community. Then, impelled by the Uranus principle, one can begin to break away from these patterns by going one's own way in life, as one can also begin to break out of the limitations of the ego structure. The challenge of consciously coming to terms with the Neptunian dimension of experience represents, potentially, a further developmental transition as it relates to the transcendence of the ego, to the experience of the numinous world of the collective unconscious, and to establishing a living relationship to the divine, in whatever form this is conceived. Finally, under the influence of Pluto, comes the possibility of the radical transformation of the structure of the psyche through the catharsis and transformation of the instinctual power of the unconscious ground.[7]

Of course, all these planetary archetypal principles are simultaneously active in many different ways across all dimensions of human experience and we should not conceive of individuation as a linear, sequential process. Rather, the course of individuation is perhaps better represented as a spiralling path or, as Jung described it, as a circumambulation around an unknown centre, one that is impelled by the mutual interaction of all the archetypal principles. If we accept, however, that in terms of the conscious realization and integration of the highest potentiality of these planetary archetypes something like this developmental progression occurs during individuation and spiritual evolution then we can consider how Eris might fit in with this scheme. Once again, in keeping with our first principle connecting interior depth and exterior expanse, we find that just as Eris is positioned beyond Pluto, farther out in the solar system, so the principle we have tentatively ascribed to the archetypal Eris – that of discord related to the tension of opposites – is only to be consciously realized during individuation *after* the descent into the Plutonic realm of the underworld. In Jungian terms, it is only after the confrontation with the repressed life power in the shadow (relating to Pluto) and the differentiation of the emotions and instincts (the differentiation of the anima, the portal to the unconscious), that the problem of opposites fully presents itself. That is to say, through individuation the tension between opposites inherent in nature is to

be realized in consciousness. And it is by means of a veritable cru-
cifixion between the seemingly irreconcilable opposites leading to
the transformation of both the ego and the Self that, according to
Jung, the Self is to be consciously actualized and made known. Again
this is a complex subject that merits a more detailed treatment,
but what is important to note here is that during the individuation process the previously unconscious or implicit discord in
existence is brought out into open. It is experienced consciously
which, although coming at the price of immense personal suffering, brings to an end the unconscious pattern of merely reacting and
responding to this discord.

In the encounter with the opposites during the individuation
process, dream and fantasy sequences become populated with imagery
representing the opposites, such as black and white, good and evil,
male and female, old and young, above and below. With each descent
the ego makes into the underworld of the unconscious, and with each
ensuing death-rebirth experience (of which there are many during
individuation), there occurs a 'passing through the opposites' to a
new world beyond, to a fuller reality. During these episodes, which
mark the death throes of the old ego, everything is experienced as
acutely discordant, as piercingly violating. Through this process, the
existential schism between ego-consciousness and the unconscious is
gradually closed, and the conscious ego comes into dynamic balance
with the unconscious. Ultimately, the opposites are reconciled in the
realization of Self, which is a *unio oppositorum* – a union of opposites
within a containing totality.

Collective forms of expression and responses

As I see it, the confrontation with the problem of opposites during
individuation represents an ideal psychospiritual response, for a
minority of people, to the evolutionary challenge set before us with
the discovery of Eris. Collectively, however, given that it is highly
unlikely that this type of individuation will happen on a wide scale
in the immediate future, we must brace ourselves, I feel, for the more
unconscious and therefore potentially problematic and destructive
forms of manifestation of the Eris archetype. Experience tells us,

looking back to the discovery of Pluto for instance, that the first appearance of a newly emergent principle often announces itself to us externally and with dramatic effect. 'When an inner situation is not made conscious, it happens outside, as fate,' as Jung said.[8] If so, then as long as the discord we have associated with Eris is not fully experienced within, as long as human beings have not assimilated their experience of the opposites into the greater identity of the Self, then we will continue to see the problematic external consequences of this discord acted out on the stage of world history. We will perforce be subject to a counterbalancing reaction to the experience of violation – both of nature and the wholeness of the human psyche – that has accompanied the development of civilization, culture, and the emergence of the modern self in recent centuries.

Pragmatically, then, if this thesis is correct, we must ask ourselves what are the possible responses to the emergence of the Eris principle? At least three distinct responses are possible. The first would see a further expansion of technological development leading to an even greater separation of the human world from nature – a further attempt to respond, largely unconsciously, to the discord underlying civilization through the increasing technological control of all natural processes. The second would see the breaking apart of civilization and the psychological structures this supports, even against our collective will, so that the strife and discord arising from nature's counterbalancing reactions become manifest in conscious experience. The first strategy, if successful, would lead at best to some form of technological utopia divorced from nature, perhaps increasing the tension and schism between civilization and the natural world, between human consciousness and its unconscious foundations. The second response might see some form of collapse, or radical restructuring of civilization and rebuilding on new foundations, as the counterbalancing reaction comes upon us externally in the form of seemingly random and arbitrary catastrophes. Of course these two responses are not mutually exclusive and they might well occur in unison. My sense, for what it's worth, is that although we might collectively strive for the former, we will ultimately experience the latter, at least to a certain extent. It remains an open question as to how far-reaching the breaking apart of civilization might be.

A third response, a middle way, if you like, is to strive for some kind of harmony and balance, embracing technological innovation while attempting to bring human civilization more into alignment with the natural order, both within and without, hopefully mitigating the deleterious consequences of the ecological crisis. Yet even with the best of intentions, and even if the optimum path were taken from this moment on, I suspect that some form of destructive counterbalancing reaction against civilization is unavoidable – indeed we have already experienced many events that seem to reflect this pattern. If so, it might well be that the challenge represented by Eris is to learn to withstand, accept, and affirm the inevitable reactionary consequences of the violation of nature during the emergence of civilization, seeking as far as possible to mitigate a catastrophic backlash that would surely follow if, collectively, a belligerent unconsciousness of our actions were to continue.

4. Mythic Associations with the Planet's Name

It is clear, when considering the qualities and themes of the existing planetary archetypes in astrology, that the character of the Greco-Roman god or goddess after which the corresponding planet is named is reflected, to a certain extent, in the archetypal meaning of that planet. Thus, to give the two most obvious examples, we find that the astrological Venus, relating to the experience of beauty, pleasure, and romantic love, closely resembles the Roman goddess of the same name (and the Greek deity Aphrodite). So too the planetary archetype Mars, like the Roman god Mars and Greek god Ares, rules war, aggression, courage, and strength. For the ancient planets, the Greek gods and goddesses were combined, in a process of mythic syncretism, with the earlier Mesopotamian and Egyptian planetary deities. Regarding the modern planets, however, we can be sure that no such syncretism was behind the naming of these planets; indeed, these planets were obviously named with no knowledge of the planet's astrological significance. It is all the more surprising, then, that the archetypal meaning of Pluto is clearly suggested by the Roman god Pluto's (and the Greek equivalent Hades') connection with the mythological underworld – all the more surprising, that is, when any mythic figure could conceivably have been chosen as the name for the planet and, as far as we know, no consideration was given to the mythological relevance of the name Pluto. With Neptune, named after the ancient Roman sea god and his Greek counterpart Poseidon, we find again that this name is suggestive of the astrological Neptune's symbolic association with the sea and with the 'oceanic' oneness characteristic of the Neptunian dimension of experience. Of course, there is much in the character of Pluto and Neptune that is not conveyed or is even contradicted by these mythological parallels, but, all the same, a readily discernible association remains.

Uranus and Prometheus

One seeming anomaly to this correspondence is Uranus whose archetypal meaning, as Richard Tarnas observes in *Prometheus the Awakener*, seems to bear only a minor connection to the mythological character of the Greco-Roman sky-god Ouranos, after which it was named, and seems instead to be more clearly associated with the deeds and character of the titan Prometheus, famed for his acts of rebellion, and for the creative brilliance, audacity, and innovation by which he stole the secret of fire from the gods as a gift to humankind. Tarnas notes that unlike the rebellious Prometheus, the god Ouranos, who was castrated and dethroned by his son Chronos (Saturn), was the victim of an act of rebellious uprising, not the perpetrator.[1] Whereas the *astrological* Uranus is rebellious and revolutionary, associated with the overthrow and disruption of the old established order, the *god* Ouranos displays quite the opposite characteristics. There seems to be no clear association, then, between Ouranos the god and Uranus the archetypal principle. If so, the general principle of drawing upon mythic associations with a planet's name to determine its archetypal meaning is, of course, invalidated.

However, a solution to this discrepancy can be found, I believe, if we consider again the two movements occurring during the individuation process. Recall that the symbolic meaning of the planetary order of the solar system can be interpreted from two directions:

1. A movement from preconscious identity with nature, from the state of *participation mystique* and animal unconsciousness, toward individual selfhood. In astrological terms this can be seen as a symbolic movement from Eris, Pluto, and Neptune towards the Sun. The outer planets here relate to the ground of being and the Sun represents the developmental attainment of ego-consciousness as it becomes differentiated from the ground.

2. A movement from the state of mature individual ego-consciousness towards the origin and ground effecting

a further transformation of the ego and the progressive realization of the Self. In astrological terms, this represents a symbolic movement from the Sun (individual ego-consciousness) towards Neptune, Pluto, and Eris (the ground).

Now the Promethean myth appears to relate to both these movements. On the one hand, regarding the first movement, it is the Promethean impulse that drives the liberation of human consciousness from its initial absorption in its containing matrix in the ground of being, winning through to attain freedom of personal will and conscious self-determination. The secret of fire stolen from the gods symbolically suggests the evolutionary attainment of the light of self-reflective awareness.

On the other hand, the myth of Prometheus also well describes the contemporary human experience of the Uranus archetype from the perspective of the second movement. Here, impelled by the Promethean impulse, the individual self in rebellion and revolt overthrows the Saturnian authority (the established order, the inhibitive morality, the oppressive superego, the rules and conventions), and goes beyond the known boundaries and limitations of the culture to bring forth its own creative expression, to individuate, to become a unique individual, and to break beyond the limits of the ordinary ego. The self pushes beyond the Saturnian boundaries to gain inspiration and illumination from the transpersonal unconscious, to further realize its freedom and uniqueness, and to bring forth its own creative genius.

Considering the myth of the castration of Ouranos, however, this is also relevant to understanding the Uranus archetypal principle in astrology, especially, I believe, to the dynamics of the first movement. From a Jungian perspective, the Ouranos myth, it should be kept in mind, although ostensibly about cosmological creation, is to be interpreted in psychological terms as relating to the emergence and development of consciousness.[2] Seen in this context, the myth of Ouranos describes what happens to each of us during the course of psychological development, when the Uranus principle of unbridled freedom, characteristic of the youth and adolescence of both individuals and the human race as a whole, is effectively castrated, as it were, by the civilized order of the social world, by the practical necessities of living, by the laws and morality of one's social group.

Instinctual freedom is brought to an end by the Saturnian principle (that is, Chronos) and, as a result, there occurs, according to the myth, a creative impregnation of the sea and the earth. In one variation of the Ouranos myth, the goddess Aphrodite 'sprang from the foam that gathered around the genitals of Ouranos, when Chronos threw them into the sea'; and in another telling, the blood flowing forth from Ouranos' wound impregnated Mother Earth.[3] Either way, what is being suggested here, I believe, is that through its encounter with the limitation and restraint imposed by the Saturnian order, the Uranus principle is elevated from a lower mode of expression (instinctual freedom but devoid of any structure or purpose) to a higher one (creative inspiration, innovation, brilliance and so forth, now directed to useful ends within the social order). Freud's analysis of the process of civilization seems to support this hypothesis: 'The replacement of the power of the individual by the power of the community,' he observes, 'constitutes the decisive step of civilization. The essence of it lies in the fact that the members of a community restrict themselves in their possibilities of satisfaction, whereas the individual knows no such limitations.'[4] The freedom and creative power of the individual (relating to Uranus), in other words, is restricted, restrained, and denied by the community (relating to Saturn).

Considering the Ouranos myth more closely, then, we can see that although, as Tarnas points out, the archetypal principle Uranus does not resemble the *character* of the god Ouranos, the myth featuring Ouranos is still relevant to understanding how the astrological archetypal principle manifests in human experience, in that the myth illuminates the psychodynamics of the relationship of the planetary archetypes Uranus and Saturn. For this reason, the hypothesis of a connection between the names of the planets and their mythic associations remains valid.

The example of Uranus, however, alerts us to an important caveat to be kept in mind when considering possible mythological parallels with astrological archetypal principles of the same name. It is clear from our knowledge of the other planetary archetypes that the principles and themes associated with them are described not only by their mythic namesakes, but by many other gods and goddesses, even where there exists no similarity in name, no etymological connection. The extent to which the mythic story of a god or goddess is directly relevant to its archetypal meaning varies. We should expect, then,

in any consideration of Eris that the myths featuring the goddess
will relate at best only in part to the archetypal Eris, suggesting only
certain qualities of the archetypal principle, perhaps only hinting
at its deeper archetypal meaning underlying more specific forms of
expression.[5] As in the case of the planet Uranus, other myths might
well bear stronger relationships to the archetypal meaning of the
planet Eris than the Eris myths themselves.

Greek myths featuring Eris

With this qualified understanding of the relationship between
astrological principles and their mythic counterparts in mind, turning
now to Eris, we can be grateful to the astronomers responsible for
naming the new planet for once again casting their net into the Greco-
Roman mythological tradition to find a suitable mythic parallel. In
this particular case, astronomer Mike Brown knowingly selected
the name Eris, the Greek goddess of strife, because of the discord
surrounding the discovery of the new planet – a discovery that, as we
noted earlier, brought about the highly controversial reclassification
of Pluto as a dwarf planet in line with the new standard of what now
constitutes full planetary status. Eris' moon, Dysnomia, was also
named with the mythic theme in mind, for in Greek myth Dysnomia
(meaning lawlessness) was the daughter of Eris.

We can immediately notice that the mythic connection of Eris
with strife and discord appears, on first inspection at least, to
generally support our earlier reflections as to the possible archetypal
meaning of the planet. To gain a deeper understanding of the mythic
Eris, however, let us consider more closely the deeds and character
of the goddess as recounted by Robert Graves in his comprehensive
compilation, *The Greek Myths*. This will, I hope, give a fascinating
insight into the planetary archetype's possible meaning and, should
this be relevant, provide clues as to how the principle might manifest
in human experience.

We learn first from Graves that in Greek mythology the goddess
Eris represents an amoral, impersonal principle of strife and violence
that 'never favours one city or party over another' but is seen, rather,
to be 'delighting in the slaughter of men and sacking of towns'.[6]

Evidently, then, if this myth is relevant to our thesis, the strife associated with Eris would seem to give rise to indiscriminate rage, slaughter, and destruction – a dark prospect when we consider the circumstances of our time, especially the threat posed by global terrorism and the devastating natural catastrophes associated with climate change. This mythologem also points to the *impersonal* nature of the mythic Eris, destroying indiscriminately without preference, perhaps suggesting an archetypal principle that is also impersonal in character, or one that appears incomprehensible and wholly unjust from the perspective of human morality. The archetypes associated with the outer planets, it should be noted, are all considered to be impersonal in nature, impervious to the sensitivities of human feeling, to the personal concerns of human life. It would make sense, therefore, if the archetype associated with Eris were similarly impersonal, demanding a high level of psychospiritual awareness and wisdom to come to terms with its deeper purpose and significance.

Second, the most well-known incident involving Eris in Greek mythology gives further detail and reinforces this theme of indiscriminate destruction: We read that because Eris was not invited to the wedding party of Peleus and Thetis she extracted revenge by proceeding to disrupt the celebrations with a mischievous prank. By rolling a golden apple at the feet of Hera, Athena, and Aphrodite with a note inscribed, 'To the fairest!' she initiated a disagreement between the three goddesses as to who was the intended recipient of the apple and who was indeed the fairest among them.[7] It was left to Paris, at the behest of Zeus, to preside over the Olympian beauty pageant that followed. All three goddesses attempted to bribe Paris, who was offered kingship over Asia (by Hera), wisdom and skill in battle (by Athena), and love of the world's most beautiful woman (by Aphrodite). Coerced by the bribe offered by Aphrodite, the 'Judgment of Paris' (see Figure 6) came down in her favour, and Paris was rewarded, as promised, by the affections of the incomparable Helen. The ensuing romance, in which Helen deserted her Greek home in Sparta and her husband, Agamemnon, to elope with Paris to Troy, brought the Greeks and Trojans into mortal conflict, and thus began the sequence of events that led eventually to the fateful sacking of Troy.[8] Decreed by the gods, the ultimate purpose of Troy's destruction, according to Graves, might have been not only to make

Figure 6. Peter Paul Rubens, The Judgment of Paris. *c.1638–39.*

Helen famous or to 'exalt the race of demi-gods,' but, ominously, 'to thin out the populous tribes that were oppressing the surface of Mother Earth'.[9] Again, in the context of the current ecological situation and global terrorism this might prove to be ominous indeed, perhaps suggesting, in symbolic terms, the relationship of Eris to the counterbalancing reactions we discussed earlier – reactions both of the Earth's self-regulating systems against human overpopulation and patterns of overconsumption, and aggressive reactions of excluded groups against the larger socio-political order.

The third and last main contribution of Eris in the Greek myths also presents intriguing clues as to the possible deeper significance of the Eris archetype and its possible forms of manifestation. We are told in Graves' account that to resolve a feud between Atreus and his brother-in-law, Thyestes, over their respective claims to legitimate sovereignty of the kingdom of Mycenae, Zeus, by an unprecedented act of interference with the natural order, persuaded Thyestes he should retract his claim:[10]

> Thereupon Zeus, aided by Eris, reversed the laws of Nature, which hitherto had been immutable. Helios [the Sun god], already in mid-career, wrested his chariot about and turned his horses' heads towards the dawn. The seven Pleiades, and all the other stars, retraced their courses in sympathy; and that evening, for the first and last time, the sun set in the east.[11]

We see here that the intervention of Eris is connected with such a momentous event, with a such dramatic display of Olympian power that, rather like Job's encounter with Yahweh in the Old Testament, it renders insignificant all merely human concerns and disputes. Although, as Graves points out, it might actually have been beyond the capability of Eris to effect such a reversal of the course of the Sun herself, she is without doubt the prime instigator of the circumstances and the conflict that stirs the mighty Zeus into action. The appearance of Eris initiates and heralds the great reversal of the Sun.

Interpreting the Eris myths

If we look to decipher the symbolic meaning of these three mythic episodes featuring Eris, several different themes are evident, each of which can, I believe, be related to the current events of our time. First, considering the reversal of the direction of the Sun, described above, this might suggest the prospect of something wholly unnatural occurring, something that transcends or violates the very laws of nature. Right away, we might think here of the hubris of human technological development, which, by going against the natural way of things, threatens to transform the natural processes of life beyond recognition: genetic engineering and modification of food, the human creation of synthetic living organisms, the possible emergence of the part-machine trans-human, artificial intelligence, the prospect of manipulated or automated evolution, the total control of emotions with chemical substances, the permanent link-up of human brains to computers using nanotechnology, the reversal of the aging process (recently achieved in mice), and much more, which all amounts to a rather dark technological vision of the future. Such developments emerging in our time could be seen as the most ambitious and extreme attempts thus far to deal with the painful discord of the human condition by attempting to bring the processes of nature fully under human technological control.

A second and, I believe, more compelling reading of the same myth offers a quite different interpretation. For it might be that the unnatural event, indicated by the change of direction of the Sun, refers not to something human beings do themselves, but rather to something that is forced upon us by the archetypal powers of the unconscious psyche (here personified as Zeus, Eris, Helios) – powers of which the modern mind is at best only dimly aware. In the world's mythic and esoteric traditions, the Sun is a symbol of the light of consciousness, and the eastern horizon where the Sun rises at dawn symbolically marks the birth of consciousness emerging out of the night-time darkness of the unconscious. Seen in this context, the myth describing the return of the Sun to the east, back to the dawn, might refer to a momentous *turning point* in which human beings find themselves inextricably caught up in a great movement back towards the origin, towards the source or ground of being, symbolized by dawn and the darkness that precedes it. This might imply the necessity for some kind of reverse movement in the evolutionary and cultural development of human consciousness – either a 'regression in the service of transcendence', to use transpersonal philosopher Michael Washburn's phrase for an essential transformative stage of the individuation process, else a regressive collapse into primitive barbarism and anarchy.[12] While the latter seems likely (think of New Orleans in the aftermath of hurricane Katrina), one can only hope that the discord in the human condition (symbolically suggested in the myth by the dispute between the two men, Atreus and Thyestes, over sovereignty) might be resolved, at least partially, by some form of necessary reconnection to the ground of being, to our own deeper nature, which could bring to the surface the immense, unresolved discord that exists between human civilization and the natural world, and that is inherent to the life process at all levels. One wonders, however, what degree of suffering and catastrophe might be necessary for this awareness to come about or if, indeed, such awareness is even possible at the current stage of cultural evolution.

One can find much to support this interpretation of the myth as some kind of turning point. Judged from many different perspectives we are living in a time of multiple endings, a time that will bear witness to the near simultaneous closure of many epochs: the end of modernity, the passing of 'peak oil', the ending of the current geological phase, the final phase of the 65-million-year Cenozoic Era that has lasted since the previous mass extinction that wiped out the dinosaurs, and – as defined by the astronomical precession of the equinoxes – the end of the two-thousand-year astrological age of Pisces. Accordingly, the contemporary *Zeitgeist* seems to be one of tumultuous change, of epochal transition. The belief that we are in or are entering some form of critical phase of *enantiodromia* or pivotal turning is widespread among social commentators of many different theoretical persuasions – a fact reflected in book titles and topics of recent years (David Korten's *The Great Turning* and Fritjof Capra's *The Turning Point* spring to mind). Indeed, the majority of the literature discussing the current world situation identifies our moment in history as a point of radical transition and transformation. We are deep in the *Kairos*, as Jung said – we are living in the 'right time' for a profound transformation of the unconscious psyche.[13]

We are fast approaching a tipping point in which, largely as a result of patterns of human consumption, ecological devastation will be unavoidable, if it isn't already. Human civilization has become too one-sided, too out of balance; it has become too far out of alignment with the natural order of things to the point that we are now courting disaster, in the form of devastating reactionary responses from nature, the Earth, and from the unconscious psyche, which are ultimately different faces of one and the same phenomenon.

This interpretation obviously supports our earlier reflections concerning the possible connection of Eris with the relationship between human civilization and nature. This myth featuring Eris, that is, lends support to the notion that the planetary archetype Eris might relate to some kind of compensatory, counterbalancing reaction by nature against its violation by human civilization, and to a reaction within the psyche to the violation that brought the loss of our unconscious wholeness during the development of ego-consciousness. Perhaps now, in meaningful coincidence with the discovery of Eris, we are experiencing some kind of turning around in human psychological and cultural development in order to re-establish a more harmonious and sustainable dynamic balance.

This idea of psychological reversal, of 'turning around' or *metanoia*, is most especially applicable to the individual. At the individual level, as we have already seen, the process of individuation, although supported and impelled by nature, is also, paradoxically, a 'running against' nature. Individuation is the arduous psychological process of overcoming the pull of the instincts and biological imperatives that, according to Jung, demands a 'violation of the merely natural man' – at least for a time.[14] In Washburn's view, too, 'regression in the service of transcendence' is an advanced phase of transpersonal development in which the conscious ego comes into contact with the lost potency, numinosity, and dynamism of the unconscious dynamic ground from which it emerged. In the dark and dangerous confrontation that ensues, the merely personal, culturally conditioned, natural human being dies, in a sense, and is reborn within the context of a larger, deeper identity. 'Regression' here implies a return of consciousness to its originating ground, and 'transcendence' refers to the overcoming of the autocratic control of rational ego-consciousness as a result of this encounter. Eris, our earlier reflections suggest, might relate not just to the experience of violation, but to the reaction that follows from it. It might relate to a compensatory backlash against one-sided development in any sphere of life.

To the extent that the individual psyche is organically embedded in the whole, in an *anima mundi* or cosmic psyche, the individual's own experience of psychospiritual rebirth might also help to mitigate the potentially deleterious and destructive effects of the Eris archetype, as we have defined it, for the entire Earth community. If the principle associated with Eris could be experienced within, consciously, then we might be able to avoid encountering this principle unconsciously, either by the acting out of this principle in the world or through events that are thrust upon us from without. In the coming struggle for the survival of the human species and the future well-being of the Earth, it is the individual with the capacity for conscious self-awareness, as Jung said, that might yet be 'the makeweight that tips the scales'.[15]

Our third theme seems to point more explicitly and concretely to the deepening problem of *human overpopulation*. According to the futurist thinker James Martin, due to the predicted exponential rise in world population, especially in India and China, together with dwindling resources of oil, water, and staple foods, we are poised within the space of a few decades to enter a perilous and potentially catastrophic global situation that he envisages as a 'canyon,' a tight passageway that, should we pass through, might deliver us to a new world on the other side.[16] It seems certain that this transition cannot occur without a vast reduction in population levels. Thus, if the myth of Eris' involvement in the fall of Troy is relevant here, the discovery of the planet Eris might prove to be a prophetic warning of what might be to come: a period of immense strife created by ecological conditions and/or terrorism whose grave consequence is 'to thin out the populous tribes that were oppressing the surface of Mother Earth'.[17] The release in 2004 of Wolfgang Petersen's cinematic rendering of the fall of Troy, in synchronistic parallel with the discovery of Eris has, in timely fashion, refreshed our memory of this myth.

Fourth, recalling that the goddess Eris wreaks havoc because she is excluded from the Olympian gathering, it might be, as suggested earlier, that Eris is a personification of something that is excluded from human affairs – something that is unacknowledged, and potentially destructive. Looking for contemporary parallels, one thinks here of the so-called rogue nations, those countries excluded from international politics, whose exclusion, although well justified, might prove to have catastrophic consequences if this is accompanied by the proliferation of nuclear, chemical, and biological weapons in line with current trends.

In psychological terms, the 'excluded' might refer to the encounter with the 'other', namely with those dimensions of our experience that are alien to our nature, and do not form part of our personal identity or that run contrary to our own natural way of being. It is the encounter with the 'other', both within, in the unconscious psyche, and without, expressed in the ways of living and personal characteristics of other people, that brings the uniqueness of our own individual character into sharper relief. When accepted and affirmed the encounter with the other – that which is opposite to us – stimulates self-awareness, promotes the realization of a greater wholeness beyond personal identity, and impels an attunement to a greater order, beyond that of personal will. The discord from this encounter often activates the undesirable, unacceptable aspects of the psyche, which have to be borne and accepted as essential to life's balance. In its problematic form, the encounter with the other can lead to the loss of individual identity in the face of external challenges, ceaseless feuding and antagonism with other people, and resentment against that which is contrary to our natural inclinations and wishes. If these themes are relevant to the archetypal Eris, it suggests this planetary principle might be to do with the challenge of embracing both what our consciousness excludes and what our civilization excludes.

As in the myth, then, in which the goddess Eris' actions were fuelled by resentment from not being invited to the wedding party, so, in line with my earlier suggestion, the Eris archetypal principle could itself be related to the *resentment and retaliation of the excluded*, both in society and in the human psyche. Perhaps Eris is related to the part of the psyche that deeply suffers and harbours resentment, that is normally excluded from conscious awareness, and that must be faced on the journey to wholeness. At the collective level, this movement towards psychological wholeness and unity seems to be closely connected, we noted, with globalization, with the realization of the Earth as a single living entity in which we are all inextricably interconnected. In this case, as discussed in Chapter 2, the archetypal Eris might represent that which must be faced en route to the achievement of collective, global unity. Like the other transpersonal planetary archetypes, Eris will most likely relate to a dimension of experience that is encountered in the process of the reorganization of human life around a greater centre, whether this new centre is planetary or the deeper individual self.

In seeking to determine archetypal meanings associated with Eris, it is not immediately apparent, however, exactly how *discord* might be related to *resentment*, so let me say a little more here about how I conceive of the relationship between them. Discord, we have noted, is an expression of the principle of evolutionary change; it associated with the movement of life between opposites that stimulates and gives rise to the evolutionary process. Resentment, as I see it, is the deep emotional response to the changes wrought by this process. Ultimately, the evolutionary process inflicts upon each of us the painful loss of our preconscious wholeness as we grow from infancy towards adult maturity. Later, for those on the path of individuation, the same principle requires the painful overcoming of much that the old ego-centred personality held dear and with which it was identified – indeed that which constituted its very essence: the sacrifice of ego-centred autonomy and freedom, the giving up of many of one's cherished ideals and dreams, the relinquishing of the need for personal power, control, and instinctual gratification. There is painful loss associated with this process, often the sense that change is forced upon one against one's will. This can give rise to resentment and, in turn, to a vitriolic lashing out against anything or anyone that appears to be the cause of one's suffering. In this sense, resentment might be seen as a by-product of evolution in human beings, just as the ecological crisis might be seen as the unforeseen response and reaction of the Earth to its unintended violation by human civilization. In astrological terms, resentment is also an expression of the Saturn-Pluto archetypal combination, with the instinctual power associated with Pluto empowering the judgment associated with Saturn, manifesting as damning judgment, loathing, or even disgust. The Saturnian judgment carries with it the repressed power of the unconscious, permitting the unconscious expression of frustrated instinct. However, the fundamental conditions in which resentment can arise, and the challenge of coming to terms with it and overcoming it, also relates to the encounter with the balance between opposites, which, if our thesis is correct, might be connected to Eris. Resentment arises because certain experiences go against the human concept of fairness and justice, which triggers eruptions of the deep, visceral, vitriolic emotions associated with Pluto. Eris, I believe, might therefore have something to do with the experience of injustice that underlies resentment, and with the process of coming to terms with this through greater spiritual and psychological awareness. We will return to this topic in the next chapter.

Our fifth point concerns the *indiscriminate rage* associated with the goddess Eris in the Greek myths, which is perhaps itself an expression of this resentment, an unconscious reactionary response to the loss of our preconscious wholeness, innocence, and natural grace during the process of psychological maturation. This rage is often unconsciously projected onto society, or certain groups that share different values to our own, and who are incorrectly identified as the cause of our pain. In terms of the collective acting out of this rage, one immediately thinks of the indiscriminate murder and destruction resulting from suicide bombings, and what seemed to many to be the equally indiscriminate 'shock and awe' tactics employed by the US military and its allies in Afghanistan and Iraq. The increase in natural disasters resulting from global warming are, alas, similarly indiscriminate in their effects, although clearly in this case we are not dealing with human rage but perhaps, as suggested above, with the 'wrath' of nature herself. The title of James Lovelock's recent book, *The Revenge of Gaia* (2006), captures this sentiment well. Lovelock's view is summarized by ecologist Stephan Harding. Drawing together several of the themes we have considered, and confirming the darkest fears of many, Harding makes clear the likely response of the planet to damaging patterns of human economic and industrial development:

> The answer will almost certainly be abrupt, catastrophic climate change, which will increase global temperatures to levels not seen for at least 55 million years. The destruction of New Orleans by hurricane Katrina in September 2005, as well as many other recent serious climatic events, are a sign that we have unleashed Gaia's wrath, and that in her desperation she seems poised to respond to our onslaught with an even greater one of her own which will kill vast numbers of people and lay low our so-called civilization.[18]

The catastrophic consequences of the 'unleashed wrath of Gaia', if it can be described that way, were starkly in evidence during 2011 with the earthquake-tsunami in Japan that opened a veritable Pandora's Box of uncontrollable hazards and threats in the form of the Fukushima nuclear crisis. On this note, it is worth considering here that, according to the Greek poet Hesiod, Eris was the mother of the *Kakodaimones* – the demonic afflictions famously released from Pandora's Box that have blighted human existence – including *Ponos* (toil), *Lethe* (forgetfulness), *Limos* (starvation), *Algea* (pains), *Hysminai* (fights), *Phonoi* (murders), *Androktasiai* (manslaughters), *Nekiea* (quarrels), *Pseuo-logi* (lies), *Amphilogia* (disputes), *Ate* (ruin), and *Dysnomia* (lawlessness).[19] Assuming that this aspect of the mythology of Eris does relate in some way to the planet's archetypal meaning, the afflictions associated with Eris' children perhaps give us an insight into the type of experiences that Eris can 'give birth to'; it is perhaps an indication, that is, of the possible range of archetypal forms of expression of the Eris principle in its destructive aspect. Although this is not a happy prospect, to say the least, let us take some consolation from the fact that the one child remaining in Pandora's Box, and who is therefore presumably yet to emerge or that remains under our conscious control, is *Elpis* (hope).

Those readers familiar with astrology will immediately recognize that all these types of 'affliction' clearly relate to various combinations of the other, already-established planetary archetypes: murderous rage, for instance, relates primarily to Mars-Pluto, lawlessness to Uranus-Pluto, ruin to Saturn-Pluto, quarrels to Mercury-Mars, and lies to Neptunian alignments, such as Mercury-Neptune or Saturn-Neptune. If so, then the association of Eris with these behaviour patterns might seem superfluous. It is perhaps the case, however, that there is something in the nature of the Eris archetype that underlies the more destructive, undesirable forms of expression of the other archetypal principles. If Eris is indeed a deep-rooted and largely unconscious principle of discord, then it might be that its archetypal influence is so profound, causing such momentous change, that it often activates the worst qualities in human nature as a desperate reaction to this change. For example, many acts of murderous rage, of extreme aggression, relate to the Mars-Pluto archetypal complex. However, it might be, given the points we have considered in the previous chapters, that Eris relates to the experiences of violation and counterbalancing reactions that underlie these aggressive actions. Pluto is instinctual force, elemental power, intensity, compulsion, extremity, and so forth, and Eris could be the experience of momentous change and reversal that both creates and calls forth this Plutonic power. There can only be a violent reaction fuelled by the repressed power of the unconscious because of the principle that creates the separation of ego-consciousness from the unconscious in the first place, because of the inherent discord in the nature of things.[20]

Another discernible theme in the Eris myths is that of *reaction against the superficial in human life*, suggested by the farcical Olympian beauty pageant that led, we noted, to the sacking and destruction of Troy. This brings to mind the superficiality that so defines our time, which we see most especially in the exaggerated worship of the body image and physical appearance, and in the often inane ramblings on social networking media such as Facebook and Twitter. The emergence of the Eris principle might thus be a necessary antidote to our modern preoccupation with image and distraction, a necessary response to superficiality, pulling our attention back to real life, back to the Earth, back to being itself.

Assuming this interpretation of the Eris myth is relevant to the planet's archetypal meaning, it seems entirely fitting that the planet Eris should be discovered at this time when collectively our connection to reality has become so tenuous – when our consensus 'reality' has become so dangerously detached and distant from the truth of our existential situation and the grave challenges we now face. The obsession with image, coupled with what seems like the inexorable movement of our culture into a hypermodern, technologically-driven virtual reality filled with cell phones, computers, games machines, and such like, has removed us so far from our natural being that one gets the sense that modern civilization is rushing headlong towards some kind of imminent explosion. The sense of both mind-numbing superficiality and unreality is reflected, it seems to me, in the predominant trend of modern conceptual art, which often amounts to nothing more than pretentious posturing or the substitution of shallow novelty for aesthetic quality and spiritual depth. In intellectual circles, too, the radically deconstructive emphasis of postmodernism has resulted, when taken to its logical conclusion, in endlessly irresolvable contextual analyses, an infinite regression into absurdity such that no viewpoint seems to have any kind of objective validity. In academia and the arts, as in many of the patterns of our everyday lives, human beings find themselves caught up in an increasingly unreal, delusional existence. One wonders again what level of catastrophe might be required to bring humanity, quite literally, back to its senses.

A more positive reading of this same mythologem points to the need for a *differentiation of the feminine principle*. This is suggested by the Judgment of Paris in which the three goddesses (Hera, Athena, and Aphrodite) could be interpreted as personifications of the anima. As Jung has shown, the differentiation of the anima (the lunar feeling realm, in astrological terms) is an essential part of the individuation process because the anima serves as a *mediatrix* to the deeper powers of the collective unconscious.[21] According to Jung, an undifferentiated anima is associated with crude instinctual desirousness and childish emotional affectivity. By contrast, after the conscious engagement with the emotions during individuation, a differentiated anima can then emerge as a newly acquired psychological function of intuitive discernment, one that complements the rational determinations of the conscious ego. Cultivating this feeling-based anima function can give one a guiding sense of the rightness or otherwise of possible courses of action and of behaviour patterns such that one can begin to heed more clearly the promptings of the Self, and better align oneself with what oriental philosophers called the *Tao* – the inherent flow and order of the cosmos.

Perhaps, then, the moral of the story of the Judgment of Paris, if there is one, is that our discernment of value must not be swayed by the promise of personal gain and should be based, moreover, on something more than exterior beauty. Referring back to the myth again, we might conjecture that the Eris principle stirs up conflict in order to draw our attention to the anima, to alert us to the necessity of confronting our emotions and desires, that we might make more conscious and informed value judgments rather than blunder into disaster solely through unconsciousness of the motives behind our actions.

Eris' gender

One important factor we have yet to directly consider, as we approach the end of this section, is that the mythic Eris is female, a goddess rather than a god. Although all the planetary archetypes seem to possess both masculine and feminine traits and modes of expression, the mythic Eris' gender does seem to have particular relevance in this case. In symbolic terms, the female principle has been closely associated with matter, with nature, and the Earth (for instance, Mother Earth, Mother Nature). Indeed, the words *matter* and *mother* have the same etymological root, and both are related to the word *matrix*, which means womb. The female principle is also associated, as noted above, with the realm of feeling and emotion. These considerations reinforce the sense that Eris is particularly connected with the ecological crisis, to our relationship with nature and the Earth, and to the necessity for individuation in order to come to terms with the long-repressed emotional dimension of the human psyche. To ignore these factors, considering all that we have said above, might provoke a violent reaction on the part of the feminine principle.

To be clear, I do not believe that Eris relates to a feminine principle *per se*, but to the counterbalancing or reactionary force of that which has been repressed, which, in modern patriarchal culture, happens to be the feminine principle, among other things. Pluto relates to the repressed power in the unconscious, but Eris, I believe, could relate to the compensatory reactive principle itself that underlies the expression of the Plutonic power.

Figure 7. The Goddess Eris. Athenian painting c.575–525 BC

The relevance of this theme was supported by the intuitive reflections of several of my colleagues at the time of Eris' naming who, no doubt influenced by the myth, felt that Eris might symbolize, or be personified by, some kind of warrior goddess. It is noteworthy here as well that from the time of its discovery until its official naming as Eris, the new dwarf planet was given the nickname 'Xena' after the warrior princess from the television show of the same name that aired at the time of the planet's discovery. For my own part, at the time I was first contemplating the possible meaning of Eris and the concurrent demotion of Pluto, I had an instructive dream that shaped my own thinking on this issue. In the dream, a powerful 'superman' figure (a godlike, heroic, instinctually empowered being that seemed to personify the planetary archetype Pluto) was confronted and opposed by a dark, almost black, goddess whose power, it became apparent, was far greater than that of the Plutonic superman.[22] I thus tentatively concluded that Eris might be a personification of the dark spirit in matter, of the spirit within the chthonic force of nature encountered after the Plutonic descent into the underworld, which needs to be assimilated into human experience.[23] I wondered whether the strife and discord inherent within nature is now calling our attention to the neglected material realm, whose innate spiritual essence has been forgotten. Perhaps, in agreement with the alchemical myths, we are now being called upon to release the slumbering spirit trapped within matter, to effect a reunion of the spirit in matter with its opposite pole: the transcendent divine.[24]

Can we then, to summarize this section, discern something in common across all the interpretations of these myths, something pointing back to a root archetypal principle? Although the Eris myths we have considered provide different perspectives from which to understand the possible nature of the Eris archetypal principle, they all, I think, can be legitimately connected to our earlier speculations that Eris, in astrology, might be related to the compensatory reaction to the growth of human civilization and its violation of nature and the Earth; to a reaction against ego-consciousness and its repression of the emotional-instinctual sphere; to reactions against patriarchy, capitalism, against technology, and against the superficiality of contemporary culture. All life involves a violation and a counterbalancing reaction, and Eris, our reasoning suggests, might relate to this process. These speculations imply that the archetypal principle associated with Eris relates to the need for restoring balance through the attainment of a higher level of unity and integration. For it is the experience of discord or strife in life that prompts the emergence of this unity through creative evolution. And if the reaction, the discord, the strife, can be experienced within individual human consciousness, this might prevent the unconscious manifestation of a destructive compensatory force through acts of nature, human violence, and through catastrophe.

5. Sources and Parallels in Philosophy, Religion, History, and Science

The analysis of the Eris myths gives us much to ponder and suggests that the archetypal Eris, to the extent that it is accurately reflected in myth, is to have a pivotal role in contemporary global affairs. It could still be objected, however, that if Eris is indeed a fundamental archetypal principle then it should be reasonably expected to occupy a more central position in Greek myth and culture than the rather peripheral, although important, interventions described above. And, sure enough, if we consider not only the mythic Eris, but also the more general principle of strife or discord, we find that this plays a highly significant, even critical, role in later Greek mythic and early philosophical thought. Strife, as Heraclitus said, is 'the father of all things'. Without strife there would be no life energy, no change, and nothing could ever happen. It is in this sense, as noted earlier, that strife or discord seems to be the motive force behind evolution, the principle that lies behind the emergence and development of human consciousness and civilization. For Empedocles, too, strife is a fundamental cosmological principle – that of separation, evil, diffusion, scattering, and division that breaks things apart and it is therefore the antithesis of love, which Empedocles conceives as the opposing current to strife.[1] Indeed, strife is necessary to draw forth love, to enable us to know love through its absence. Thus, if strife is now coming to the foreground of human collective experience, its effects could perhaps be alleviated if we are able to manifest, as best we can, its compensatory opposite.

Strife and justice in Greek philosophical speculation

A number of speculations on the nature of the astrological Eris have connected the dwarf planet to the principle of chaos, as an extension, no doubt, of the mythic association with strife or discord. The relationship between strife and chaos, however, is not as straightforward as it first appears. In Greek mythic and early philosophical speculation, in which the principles of chaos and strife both play a significant role, the word *chaos* has a very different meaning to what we commonly understand by the term, as F.M. Cornford explains: '"Chaos" was not at first, as we conceive it, formless disorder. The word simply means the "yawning gap".'[2] This 'yawning gap' points to the initial separation of earth from sky, which is effected by strife. For instance, in the Orphic cosmology we learn 'how earth and sky and sea were at first joined together in one form, and then disported, each from each, by grievous strife'.[3] Cornford suggests that this account of the separation by strife in cosmic creation mirrors accounts in Babylonian myth, primitive Egyptian myth, Judaism, and Taoism in which chaos or some primal unity is initially divided into opposing substances. The 'yawning gap' calls to mind our earlier connection of Eris with pairs of opposites, especially the separation of human consciousness and human civilization from nature.

   The principle of strife also plays a central role in the cosmologies and accounts of creation of Heraclitus, Empedocles, and Anaximander. According to Cornford:

> The coming into existence of individual things is variously
> attributed by the early cosmologists to love or harmony,
> and to feud, strife, or war. The two representations are, as
> Heraclitus insisted, not so irreconcilable as they seem to be at
> first sight. They are only two ways of conceiving the meeting
> of contraries. The two contraries are antagonistic, at perpetual
> war with each other.[4]

According to the accounts of these early Greek thinkers, during the process of cosmic creation the original unity of the primordial One was divided into opposing contraries, into pairs of opposite qualities and elements. 'Between the two members of each pair of contraries,' Cornford explains, 'there is antagonism, strife, feud ... Out of this strife ... arises, according to Anaximander, all individual existence, which is the offspring of war and injustice.'[5] Cornford notes that in Anaximander's cosmological scheme, 'The war of antagonistic principles ... generates the whole world of things we see.'[6] 'In other systems,' he continues,

> not only all existence, but all goodness and perfection in the visible world, involve a balance or harmony of opposed powers – a reconciliation in which the claims of both are, if only temporarily, adjusted. Besides War, there is also Peace; besides Hatred and Feud (*Neikos*, *Eris*, etc.), there is also Love and Agreement (*Phila*, *Harmonia*, etc.). This general scheme of conception runs through all ancient physical speculation, and ... all ethical speculation also.[7]

It is the function of the interaction of the contraries, love and strife, to bring separate individuals, groups, and elements into a greater union and harmony. Applied to social groupings, as Cornford explains:

> The Love which draws all the elements into the indiscriminate mass, called the Sphere, corresponds to the solidarity of the whole tribe [or society or civilization]. Strife, or feud, is the disintegrating force, which causes segmentation into minor groups.[8]

Such fragmentation is, of course, characteristic of the postmodern era (in the pervasiveness of splinter groups and national or political independence movements, for instance, and with the rise of ever-greater specialisms in academia and the professions), yet this fragmentation has been simultaneously accompanied by moves to form greater political and economic wholes (through organizations such as NATO, the UN, the European Union), and recently through global alliances to combat terrorism and climate change.

Heraclitus relates strife to the harmony between opposites, and to the principle of *enantiodromia* – the tendency of one quality or principle to bring forth its opposite when taken to its extreme –

reinforcing the notion that Eris might be related to turning points. Crucially, Heraclitus also connects strife with the cosmic principle of *justice*.

Cornford draws out Heraclitus' position in an imagined debate between Heraclitus and Anaximander concerning the relationship of strife and justice. According to Cornford, Heraclitus would have said something like the following:

> You admit ... that 'War is the father of all things' (frag.44), and yet you condemn the parent of all life as unjust. The end of warfare would be the end of life itself. 'Homer was wrong when he said: "Would that Strife might perish from among Gods and men!" He did not see that he was praying for the destruction of everything; for, if his prayer were heard, all things would pass away'. (frag. 43) Death is not dissolution, but rebirth; so, war is not destruction but regeneration. 'War is common to all, and Strife is Justice, and all things come into being through Strife.' Strife is Justice; if it were not for these acts of 'injustice,' as you call them, men would not have known the name of Justice. Justice is not the separation of opposites, but their meeting in attunement of 'harmony.' Without opposition there were no agreement. 'What is at variance agrees with itself. It is the attunement of opposite tensions, like that of the bow and the lyre.' (frag.45)[9]

In another telling passage, Cornford follows (and cites) Heraclitus in his association of justice with the law of the One or the All:

> This Justice or Harmony, again, is the Logos, the Spirit of Life, observing measure, but passing all barriers. It is the divine soul-substance, whose life consists in movement and change. It is also the one divine Law, the law of Nature (*physis*), which is the Will of God. 'It is Law (*nomos*) to obey the will of One.' (frag. 110) This is true for the universe, no less than for human society; it is common to all things. 'Those who speak with understanding must hold fast to what is common to all, as a city holds fast to its law, and even more strongly. For all human laws are fed by the one divine law. It prevails as much as it will, and suffices for all things with

something to spare.' (frag. 91b) 'So we must follow what is
common; yet the many live as if they had a wisdom of their
own.' (frag. 92) 'It is not meet to act and speak like men
asleep. The waking have one common world, but the sleeping
turn aside each into a world of his own.' (frag. 94, 95)[10]

On reflection, it would perhaps be more accurate to say that human
laws (relating to the Saturn archetype in astrology) *should* be fed
by the one divine law for, clearly, the gulf between these can be
considerable. The justice implied here in Cornford's discussion of
Heraclitus is not human justice or anything to do with individual
rights or even human rights; it is *divine* or *cosmic* justice. And,
needless to say, human and divine justice are often very different from
each other, since the law of the One is barely comprehensible in terms
of the norms and strictures of human morality.

Given what we have said before about the possible correlation of
the Eris archetypal principle with counterbalancing reactions arising
from opposites inherent in the nature of things, the association
of Eris with the principle of justice makes good sense. We can see
that in broad coincidence with the discovery of Eris, collectively we
now appear to face the challenge of coming to terms with the gross
injustice of both atrocities perpetrated by terrorists and natural
catastrophes related to ecological and climate changes – phenomena
which might, however, reflect the scarcely intelligible workings of a
deeper principle of cosmic order and an impersonal cosmic justice by
which all life is held in dynamic balance.

Heraclitus alludes to the inherent conflict between the will of the
One and the many 'who live as if they had a wisdom of their own'
and 'act and speak like men asleep'. He thus associates justice with the
law of the One, of the All, rather than the will of the collective 'herd'
or the individual will. It is perhaps the case, if the Eris archetypal
principle is indeed related to cosmic justice, that any deviations from
the workings of this principle manifest as discord and strife. In other
words, strife arises when human will and actions unavoidably enter a
state of disharmony with the will of the One. Strife, by this account,
is the expression of conflict with a divine or cosmic justice. With the
development of ego-consciousness and individual human autonomy,
the will of the individual and of human society have become ever more
discrepant from the law of the One. It is the work of individuation,

as Jung describes, to bring the individual will into alignment with
the will of the whole, traditionally conceived as the Will of God – a
process that moves forward through experiences of painful discord.

What is important here is that Heraclitus connects strife and
discord with wholeness, thereby lending support to our suggestion
in earlier chapters that the Eris archetypal principle is connected
with a movement through discord towards greater planetary unity
and wholeness. Rather than just fuelling a blindly driven will, as
Schopenhauer thought, discord seems to have a positive purpose
within the evolution of the modern world, even though this is often
incredibly hard to discern in the midst of the trials and tribulations,
and crises and catastrophes, of human life.

The principle of justice, then, impels human beings to come
into alignment with the dictates of the greater planetary and
universal wholes of which they are a part, and it is also related to
the reconciliation of the opposites. Thus Cornford: 'When once
we understand that Justice is the Way of Life, and also the force
that moves along that way and owns no barriers, the doctrine of the
harmony of the opposites falls into line.'[11] It is through the repeated
experience of acutely painful discord that one learns to recognize and
consent to a higher order, a divine will, beyond the opposites of good
and evil, or pleasure and pain, or right and wrong. The experience of
strife is thus simultaneously an impulse towards the realization of a
higher form of justice, harmony, and wholeness.

Justice and the balancing of nature in aikido

Looking outside the Greek tradition, we find that a similar perspective
on justice and harmony in the philosophy of the Japanese martial art
aikido. As Mitsugi Saotome explains in his Aikido and the Harmony
of Nature, the concept of harmony, seen from a cosmological or
spiritual perspective, necessarily encompasses strife, conflict, and
discord. The principle of universal harmony and justice, he suggests,
is often at variance with normal standards of human fairness:

> We sharply separate harmony [as pleasant and good] and
> conflict as harsh and evil; with each person standing in the

center of his or her own private universe, we decide what
is fair from the limited perspective of one individual ...
We do not comprehend the all-encompassing harmony of
nature, nor truly understand the exquisite justice of the
Creator's universal laws. Harmony does not mean there are
no conflicts, for the dynamic spiral of existence embraces
both extremes. Conflict is the beginning of harmony as
death is the beginning of life. Filled with conflict, the flow
of nature remains flexible enough to adapt and change,
bringing all of creation back into balance.[12]

This principle of dynamic natural harmony – of homeostasis, as it
is also known – is thus to be conceived as an impersonal force that
unswervingly strives to maintain a balance within nature as a whole.
Obviously such a principle has no regard for the personal concerns
of the individual. From this perspective, natural disasters, as an
expression of the irreversible climate change we are now experiencing,
reflect nature's attempts to maintain balance in the face of the
imbalance created by human 'progress' and development. Alas, the
workings of this balancing principle are often violent, destructive,
and incomprehensible. Saotome adds:

Labeled as destructive violence by the narrow vision
of humanity's ego, the natural phenomena of volcano,
earthquake, tidal wave, and typhoon are nature's immaculate
system of adjustment, the mighty force of harmony.[13]

'Nature's laws are harsh without exception', he observes, and
continually draw forth evolutionary adaptation and adjustment. 'If
a life form, any life form, cannot change and adapt, it will perish, for
the balancing of nature will continue.'[14] Thus, if Eris is related to a
cosmic balancing principle, it must play a crucial role in evolution,
creating the circumstances that continually draw forth evolutionary
adaptations and growth at all levels of life.[15]

From a metaphysical perspective, the workings of a principle of
cosmic balance are also described by the Indian notion of karma,
relating to actions and reactions that govern all of life, even across
different incarnations. At a psychological level, too, Jung's discovery
that the unconscious psyche strives to maintain a state of dynamic

harmony is relevant here. The unconscious, Jung discovered, brings forth images in dreams and fantasies that serve to 'compensate' for the one-sidedness of an individual's conscious attitude. Such images present consciousness with another, counterbalancing perspective that enables the individual to attain a more rounded and complete state of awareness.[16]

The constructive value of strife

The complementary opposite principles of love and strife are also brought together in the teachings of Christ. While Jesus' proclamation of the gospel of love – his exhortation to the love of one's neighbour – is universally known, his association with strife, by contrast, has been almost entirely overlooked or downplayed. A fuller picture of a Jesus who is not only the living embodiment of love but also, where necessary, the advocate and harbinger of strife emerges from a closer inspection of the Gospel sayings. The Jesus of *The Gospel of Thomas*, for instance, in a version of a teaching also found in Matthew and Luke, is recorded as saying: 'Perhaps men think that I have come to cast peace upon the world and they do not realize I have come to cast divisions upon the earth: fire, sword, strife.'[17] Once again this seems to allude to the necessity of the acceptance and affirmation of strife and discord as essential to the highest pattern of human psychospiritual response to the challenge of life.

Friedrich Nietzsche, too, gives a more positive moral valuation of strife, arguing that contrary to the modern understanding, which sees strife as something negative, undesirable, and therefore bad, strife was originally associated with what was deemed 'good', noble, and healthy in human experience. The Latin term *bonus*, meaning good, Nietzsche argues, was originally derived from the earlier term *duonus*, which was itself associated with *bellum*, meaning war. 'Therefore *bonus* as the man of strife, of dissention (duo), as the man of war: one sees what constituted the "goodness" of a man in ancient Rome.'[18]

Nietzsche was born with Eris in major alignment (a conjunction) to Neptune (see Figure 8), and, in keeping with our speculations on the meaning of Eris, his life work marked a vitriolic reaction against many areas of life associated with the Neptune archetype: religion,

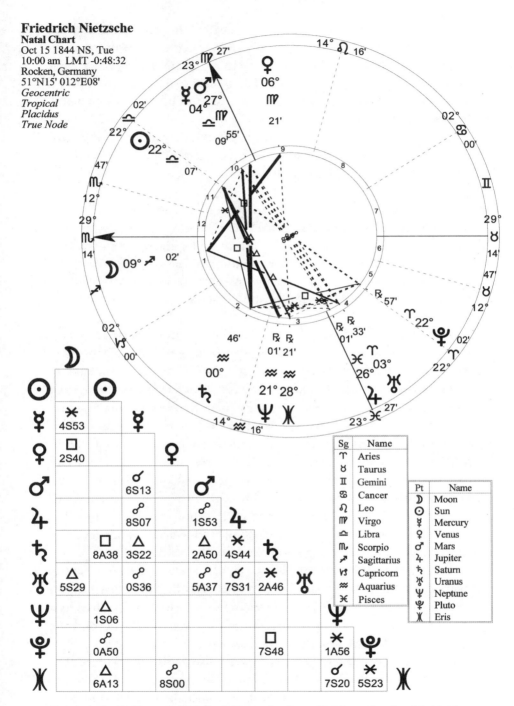

Figure 8. Birth Chart for Friedrich Nietzsche. Born 10.00am, October 15, 1844, Rocken, Germany. In the above, Eris is represented by the symbol ⚸

spirituality, metaphysics, Platonism, ideals, Christian morality, the emotion of pity. His philosophy also advocated a momentous turning point in human spiritual history in its attempt to reverse what he saw as the damaging effects Christianity had had on human spiritual development by its disparaging attitude towards natural, instinctual life.

Yet Christianity and Platonism, Nietzsche believed, also had positive values as an oppositional force that might lead to higher spiritual realization, a higher morality:

> But the fight against Plato or, to speak more clearly and for
> 'the people', the fight against the Christian-ecclesiastical pressure
> of millennia – for Christianity is Platonism for 'the people' – has
> created in Europe a magnificent tension of the spirit the like
> of which had never yet existed on earth: with so tense a bow
> that we can now shoot for the most distant goals.[19]

The reaction against Platonic metaphysics and Christianity was simultaneously a recognition of the significance of these perspectives for creating a tension of opposites that served, in turn, as a spur to excellence, to the attainment of higher goals. Platonic and Christian metaphysics, in Nietzsche's view, drew forth a compensatory reaction by setting human spirituality and reason against instinct. This reaction prompted the emergence of a higher unity that might reclaim those aspects of existence which Platonic metaphysics and Christian morality excluded, namely, the power of the instincts, the body, and the Earth. Such themes obviously fit with our speculations concerning the meaning of Eris, especially its possible connection to strife, tensions of opposites, movements towards higher levels of consciousness, and turning points and great reversals.

At the heart of Nietzsche's philosophy was the impulse to transcend mediocrity. He believed that strife and enmity are not something to be avoided, but something to be welcomed, affirmed, to be used as a spur to excellence and the realization of a greater totality beyond the opposites of pleasure and pain, good and evil.[20]

This alternative interpretation of strife also finds support from the Greek myths. According to the account of Hesiod, there were actually two forms of the goddess Eris that were eventually conflated under a single name: she is not only the destructive goddess of strife we have focused on here, but also the goddess of 'healthy competition' who promotes excellence and acts as a spur to greater achievement – an interpretation which seems consistent with our linking of Eris to the principle of creative discord that has given rise to human civilization and that informs the entire evolutionary process, drawing forth evolutionary advances through the continual challenge of adapting to ever-changing life circumstances. In Hesiod's words:

> So, after all, there was not one kind of Strife alone, but all
> over the earth there are two. As for one, a man would praise
> her when he came to understand her; but the other is
> blameworthy: and they are wholly different in nature. For
> one fosters evil war and battle, being cruel: her no man
> loves; but perforce, through the will of the deathless gods,
> men pay harsh Strife her honour due. But the other is the
> elder daughter of dark Night, and the son of Cronos who
> sits above and dwells in the aether, set her in the roots of
> the earth: and she is far kinder to men. She stirs up even
> the shiftless to toil; for a man grows eager to work when
> he considers his neighbour, a rich man who hastens to
> plough and plant and put his house in good order; and
> neighbour vies with his neighbour as he hurries after
> wealth. This Strife is wholesome for men. And potter is
> angry with potter, and craftsmen with craftsmen, and
> beggar is jealous of beggar, and minstrel of minstrel.[21]

The Indian philosopher Sri Aurobindo, to give a further example of a more positive rendering of the value of strife, thought that the experience of strife and discord, whether in worldly affairs or in the human psyche, is ultimately in the service of a greater harmony, fulfilling an evolutionary purpose:

> For all the problems of existence are essentially problems
> of harmony. They arise from the perception of an unsolved
> discord and the instinct of an undiscovered agreement of
> unity. To rest content with an unsolved discord is possible
> for the practical and more animal part of man, but
> impossible for his fully awakened mind.[22]

Again, by this account, strife or discord plays an essential role in human psychospiritual development in that it demands its own resolution; it elicits a higher level synthesis to produce a greater condition of harmony.

A similar viewpoint is proffered by the philosopher and mystic Alan Watts. 'When we adjust our lenses to watch the individual cells of an organism,' Watts wrote in *The Two Hands of God*,

> we see only particular successes and failures, victories and
> defeats in what appears to be a ruthless 'dog-eat-dog' battle.
> But when we change the level to observe the organism as
> a whole, we see that what was conflict at the lower level is
> harmony at the higher: that the health, the ongoing life of
> the organism is precisely the outcome of this microscopic
> turmoil. Now the expansion of consciousness is no other than
> extending our vision to comprehend many levels [of nature]
> at once, and, above all, to grasp those higher levels in which
> the discords of the lower levels are resolved.[23]

For Watts, the resolution of this conflict at a higher level of consciousness is inherent in a perennialist spiritual perspective. It is possible, he believed, to direct one's attention to the 'perennial intuition of the implicit concord and harmony which underlies the explicit discord and conflict of life'.[24]

Watts was born with Eris in a square (90-degree) aspect to Saturn and to Pluto. The Saturn-Pluto archetypal pairing is evident above in the descriptions of the struggle for survival in the harsh world of nature, the 'dog-eat-dog' battle, but the Eris principle, in light of the hypothesis we are considering here, might well relate to the resolution of this struggle and conflict through the realization of higher levels of consciousness or higher unities. Watts also had Eris in a 120-degree trine alignment with Neptune, which is usually considered to indicate a well-established, supportive relationship between the corresponding archetypal principles. This Eris-Neptune alignment perhaps explains, from an archetypal perspective, why he draws upon the Neptunian areas of spirituality and religion to mediate the resolution of conflict and strife through the realization of an underlying harmony.

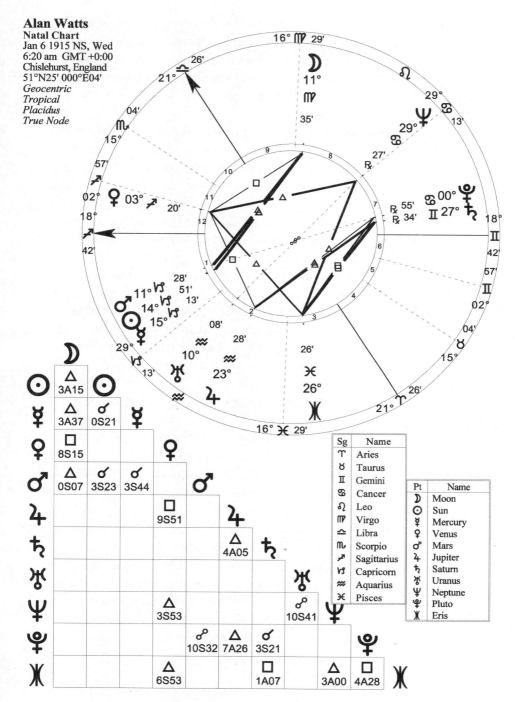

Alan Watts
Natal Chart
Jan 6 1915 NS, Wed
6:20 am GMT +0:00
Chislehurst, England
51°N25' 000°E04'
Geocentric
Tropical
Placidus
True Node

Sg	Name
♈	Aries
♉	Taurus
♊	Gemini
♋	Cancer
♌	Leo
♍	Virgo
♎	Libra
♏	Scorpio
♐	Sagittarius
♑	Capricorn
♒	Aquarius
♓	Pisces

Pt	Name
☽	Moon
☉	Sun
☿	Mercury
♀	Venus
♂	Mars
♃	Jupiter
♄	Saturn
♅	Uranus
♆	Neptune
♇	Pluto
⯰	Eris

Figure 9. Birth Chart for Alan Watts. Born 6.20am, January 6, 1913, Chislehurst, England.

Watts observes that the inner unity of the opposites is suggested by Taoist, Buddhist, and Hindu myths of the 'primordial pair' and the 'cosmic dance'.[25] In the West, the inner unity of opposites increasingly gives way to outright conflict between opposites, he notes, as in myths featuring conflict between brothers, such as those of Horus and Set, and Ohrmazd and Ahriman.

In Christian mythology, with the absolute separation of good and evil into opposing forces symbolized by Christ and Satan, respectively, all sense of an underlying unity between opposites is lost. Salvation or redemption then lies in the recovery of the lost unity. 'The human ideal becomes, then,' Watts explains, 'the hermaphroditic or androgynous sage or "divine man", whose consciousness transcends the opposites and who, therefore, knows himself to be one with the cosmos.'[26] Again, in light of our suggested meanings of Eris, these reflections are all consistent with the aspects between Eris and Pluto and Eris and Neptune in Watts' chart.

Born just a few months before Alan Watts, the ecological theologian Thomas Berry also had Eris in a trine to Neptune and a square alignment to Saturn-Pluto. Like Watts, Berry focused on the higher unity of world religions, especially within the context of the ecological crisis, which was the central concern of much of his work. Saturn-Pluto themes are very much evident in Berry's writing: the human species facing a grave challenge (a 'great work') that has been thrust upon us with the force of necessity; the inescapable labour of transformation required to come to terms with the ecological crisis; the inherent brutality and struggle of life; the immense destructive powers of nature.[27] Yet interwoven with these themes, as the following quotes indicate, was an awareness of the creative potentiality of tensions of opposites moving humanity toward the realization of a more inclusive planetary unity. 'Life emerges and advances,' Berry suggests, 'by the struggle of species for more complete life expression ... From Heraclitus to Augustine, to Nicholas of Cusa, Hegel, and Marx, to Jung, Teilhard, and Prigogine, creativity has been associated with a disequilibrium, a tension of forces.'[28] At the socio-cultural level, he observes, 'Nations exist in an abiding sequence of conflicts ... [but] the issue of inter-human tensions is secondary to earth-human tensions'.[29] Advocating a cosmology of peace, Berry declares that today 'everything depends on a *creative resolution of our present antagonisms*'.[30] 'I refer to a *creative resolution of antagonism* rather than to peace,' he explains, 'in deference to the violent aspects of the cosmological process.'[31] The ideal situation Berry finds described in the words of A.L. Kroeber: to exist at 'the highest state of tension that the organism can bear creatively'.[32] 'The peace of Earth,' Berry contends, is '... a creative process activated by polarity tensions'.[33]

Regarding the modern political situation, he adds:

> The severe tensions existing among the great [economic and political] powers are of a planetary order of magnitude because the resolution of these tensions is leading to a supreme achievement: the global unity toward which all earthly developments were implicitly directed from the beginning.[34]

As a way through these tensions and the profound transformation we are now experiencing, Berry evokes the Neptunian theme of the dream: 'the dream of the earth. Where else can we go for the guidance needed for the task that is before us?'[35]

Another figure who addressed the relationship of opposites within the context of religion and spiritual transformation was William Blake, who was born with Eris in a conjunction with Pluto, and trine to Neptune. According to M.H. Abrams in *Natural Supernaturalism*, his masterful study of Romanticism, 'Blake's prophetic books narrate various stages of the division and reintegration of the Universal Man' who undergoes a 'fall into Division' (in Blake's words) resulting in a 'fragmentation of unitary man both into isolated individuals and into an alien external world' – a 'dreadful state / Of Separation', as Blake describes it.[36] From a psychological perspective, Abrams points out, this fall describes 'a progressive dissociation of the collective human psyche into alien and conflicting parts, each of which strives for domination.'[37] To overcome the division a spiritual transformation must take place, restoring in a higher state the original condition of undivided unity. In a mystical vein, this re-union is conceived as a mythical marriage between 'the severed male and female opposites.'[38]

Abrams argues that this pattern of fall/division and resurrection/ unity is also working itself out through history: 'The dynamic of this process,' he explains, 'is the energy generated by the division of unity into separate quasi-sexual contraries which

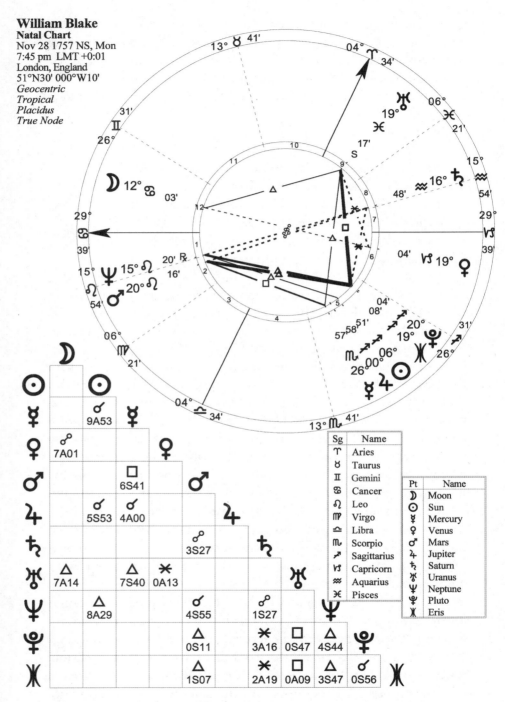

Figure 10. Birth Chart for William Blake. Born 7.45pm, November 28, 1757, London, England.

strive for closure.' In Blake's vision, the contraries are 'opposing yet complementary male-female powers, which, in their energetic love-hate relationship, are necessary to all modes of progression, organization, and creativity, or procreativity'.[39] As Blake declares, very much in the spirit of Heraclitus:

> Without contraries is no progression. Attraction and
> Repulsion, Reason and Energy, Love and Hate, are
> necessary to Human existence. From these contraries
> spring what the religious call Good & Evil. Good is the
> passive that obeys Reason. Evil is the active springing from
> Energy. Good is Heaven. Evil is Hell.[40]

These themes, of course, are remarkably consistent with our earlier speculations as to the meaning of Eris, especially its association with the separation of human experience from an original unity, and opposites as the foundation of all life energy, with the 'creative strife of contraries' acting as an impetus to evolutionary progression and psychospiritual transformation.[41] In the above passage, Blake connects these themes to religion (relating to the archetypal Neptune), pointing to the necessity of evil as a natural counterpart to good, and to hate as the requisite complement to love, thus radically challenging and running against the accepted absolute dualism of good and evil in Christianity, which depicts the instincts and energies of the body (relating to the Pluto archetype) as evil. Blake's analysis, perhaps reflecting Eris' alignment to Pluto in his birth chart, points to the possibility of a higher level perspective, beyond good and evil, that can accept and affirm both.

For Blake, as for many Romantic writers, the movement from original unity to fragmentation and separation, and then on to the attainment of a higher unity, is a fundamental pattern of human experience. As we have seen, this process moves forward through the strife and discord arising from the interchange of opposites – all themes which, if this thesis is correct, can be connected to the archetypal principle associated with Eris. Incidentally, Abrams, born July 23, 1912, also has major Eris-Pluto (a square) and Eris-Neptune (a trine) alignments in his birth chart, helping him to home in on archetypally pertinent elements of Romantic literature, in general, and Blake's own analysis, in particular.

These brief examples alert us to the higher potentialities of Eris, as we are speculatively defining it here, and help us to contrast the possible negative and positive expressions of that archetype. More problematic modes of expression of Eris might include vitriolic or indiscriminate reactions against perceived injustices; surrender in the face of seemingly irreconcilable tensions or disputes; one-sided behaviour patterns and living in ignorance of the 'other'; the potentially catastrophic consequences of living in disharmony with the natural systems of which we are a part; and living according to personal will in ignorance of divine will. A more positively expressed Eris, however, might embrace discord and rise above seeming injustice in the pursuit of excellence and the realization of psychospiritual wholeness and planetary unity. It might also be associated with bringing one's personal will into alignment with the cosmic or divine will in the realization of a higher form of cosmic justice or balance, and it might relate to using the conflict of opposites as the necessary impetus for evolutionary development.

Eris and world history

At a collective level, the working out of the same principle seems to be evident in the course of world history in which the experience of international discord continually stimulates attempts to achieve greater accord and harmony, ultimately, one hopes, forging a closer union between the nations of the world. Thomas Berry and Brian Swimme have described this process as it was played out in the late nineteenth and early twentieth century Europe.

> [The] colonial wars for the western control of non-European lands brought about continual conflict among the western nations themselves throughout the period of colonial expansion ... The never-ending strife resulted in World War I and the vain effort afterward to establish an effective League of Nations whereby the peoples of the Earth could live in peaceful relations with each other.
>   The League of Nations proved most effective in its humanitarian activities but eventually nonviable in the

political order. Antagonism between the nations was too intense. The will for peaceful settlement of disputes was almost totally lacking in the member nations. There was too much nationalist resentment between them, too much rivalry for control of territory. Political ideologies seized too violently on Russia, Germany, and Italy. Economic depression settled over the world after 1929. The times were not propitious, though the efforts made by the League of Nations were admirable.[42]

We can see, in this passage, several of the themes we have provisionally ascribed to the Eris archetypal principle: the strife, conflict, and rivalry between nations that elicit attempts at resolution through closer international relations; discord prompting a step towards global unity, albeit falteringly; and resentment as a reaction to the discord, hindering this process, and leading in this case to the further eruption of immense strife, violence, and destruction of the First World War. We can also discern how the Eris archetype seems to be distinct from, and interacts with, the expression of the other planetary archetypes associated with the outer planets, Neptune and Pluto. For example, we can see here how the drive for power and control (associated with Pluto) maintained the state of perpetual strife in Europe, with nationalist resentment fuelling, reciprocally, the Plutonic drive for power. Ultimately, this created an energetic charge in the collective European psyche that erupted in the twentieth century's two world wars, and left it susceptible to the powerful myths and political ideologies (relating to Neptune-Pluto in combination) that resulted in the Nazi and Stalinist regimes (Pluto). More positively, however, the strife and conflict also stimulated the humanitarian idealist impulses (relating to Neptune) from which major international organizations came into existence, beginning with the League of Nations at the end of the First World War.[43]

In certain respects, the First World War seems to be a prime example of many of the themes we have discussed in relation to Eris, especially the strife and discord prompting the movement towards greater planetary interconnectedness, and violent reactions and counter-reactions leading to the loss of life on a massive scale. The First World War was also, of course, an irreconcilable conflict that brought a major turning point in the history of human civilization, heralding the terminal decay of the era of 'empire' (initiated by the destruction of the Ottoman Empire) and ushering in a new era of global politics in the place of imperialism.

Something like an underlying *telos* appears to be at work in this process, one that pulls nations into relationship, whether this relationship is harmonious or conflicting – a ceaseless dialectic between feud and peace, discord and harmony, moving the peoples of the Earth, often in blind unconsciousness, towards an evolutionary goal we have not been able to discern. Perhaps, indeed, a principle of this kind has been the motive force behind the dramas of human history from the outset. From the migrations of the early humans across the plains of Africa, to the Crusades and Colonial conquests, and to the invasions and great wars that have shaped history – perhaps all these have been unconsciously manifesting a hidden purpose, working towards the realization of planetary unity or perhaps even to the manifestation of a divine will in human affairs. It is out of this process that recently the global economy has emerged; and it is through this process now that a unified planetary civilization must come into being if ecological disaster is to be avoided.

Eris and the ecological crisis

Just as 9/11 was the event that marked the beginning of the age of global terrorism, so it was the Indian Ocean tsunami of late December 2004 and the New Orleans tragedy caused by Hurricane Katrina in September 2005 that were seen, rightly or wrongly, as the most significant expressions of the damaging consequences of climate change and as a sign of what might be to come. It was these events, accompanied by many other incidents and followed up by media coverage and book and film releases, that brought the ecological crisis decisively into public awareness. Like 9/11, these two events were global in outreach, dramatic instances of the compensatory, counterbalancing reactions of nature against the deleterious effects of human civilization on the planetary biosphere.

In line with our earlier analysis of the myths of Eris, we see also that in both these cases, as with 9/11, there occurred a splitting apart of civilization, a collapse of the human world, accompanied by the eruption of immense strife, with each event serving to pull the collective attention from the superficial unreality of modern culture back to reality. Tragically, Eris' mythic connection with the destruction of Troy and the reduction of population levels might also be borne out by these crises. Over 200,000 people died as a result of the tsunami. Occurring just one month after Katrina was the Kashmir earthquake of October 2005 affecting India, Pakistan, and Afghanistan in which 73,000 people perished and three million were made homeless. This was followed in 2008 by two more major natural disasters. In Burma an estimated 78,000 people were killed by the cyclone that tore through the country in the early days of May. A further 56,000 people were unaccounted for, and 2.5 million more people were left in dire need of food, shelter, and medical aid. Just over one week later, in China, an earthquake took the lives of an estimated 70,000 people, with another 18,000 unaccounted for. In early 2010, the catastrophic earthquake in Haiti claimed the lives of over 300,000 inhabitants, and affected more than three million people in total. In a sequence of what seemed like one major disaster after another in the years close to the discovery of Eris, such catastrophes were a stark, terrifying demonstration of the fragility of the human condition and the volatile state of the planet – a condition further reinforced by recent natural disasters in Australasia and the catastrophic combination of earthquake-tsunami-nuclear crisis that befell Fukushima, Japan, during March 2011.

In many of these natural disasters, we can observe the devastating consequences of the immense counterbalancing forces of nature at work, forces indifferent to all human standards of fairness, and calling forth a global response to the ecological crisis, thus eliciting positive change and growth. As is becoming all too apparent, it often takes immense tragedies of this kind to drive home the reality of our ecological situation, a reality that can be ignored even when confronted with the most alarming facts and figures about global warming, overpopulation, rising sea levels, disappearing ice-caps and rainforests, and species becoming extinct. It often requires something dramatic to grab the collective attention and galvanize political will to action.

For the horrors of these events, one can see in retrospect how each of them has contributed, in many cases markedly, to closer global connectedness, even where the basis of this interconnection between countries remains fraught with difficulty and tension. As we have seen, history has shown that time and again the experience of acute strife and discord, in many different forms, has drawn countries out of isolation into fuller participation in world affairs (a phenomenon dubbed 'earthquake diplomacy'). Such natural disasters, we must surmise, are ultimately unavoidable. But, if these natural disasters are increasing in frequency and severity, as they seem to be, they might be taken as signs and symptoms of the damaging consequences of the way modern civilization has developed – a consequence, that is, of the violating effects of modern industrial civilization on nature. In each crisis, the gulf between the human world and nature has been starkly exposed and, in several cases, as the structures of civilization have collapsed, this gulf has been closed. Needless to say, each occurrence resulted in immense strife for the victims and their families, a sense of gross injustice, and most of the events tragically resulted in the large-scale loss of life. Finally, a number of the events (relating both to global terrorism and the ecological crisis) were in some way major turning points, events that changed the future course of civilization. In sum, many of the themes we have provisionally assigned to Eris based on analysis of Eris myths appear to have been prominent in events of recent years, coincident with the planet's discovery.[44]

Of course, we cannot rule out the possibility that any of these events might relate to archetypal themes associated with other of the newly discovered planet-like bodies in the solar system. Certainly, many of these events can be meaningfully described in terms of the archetypes associated with the existing planets in astrology. For example, Richard Tarnas has pointed out that both the hurricane Katrina and the Indian Ocean tsunami tragedies are pervaded by many themes associated with the planets Saturn and Neptune, which were in major alignment at the time: death by drowning, water permeating and sweeping away physical structures, the dire suffering and sense of destitution, the contamination of water supplies, dehydration, the lack of essential medicine and aid – together giving rise to a demoralizing 'collective sense of hopelessness and despair'.[45] Reflecting the still active Saturn-Neptune opposition, many similar circumstances and themes were also evident in Burma during 2008.

Equally, the Al-Qaeda terrorist attacks in September 2001 reflected the archetypal themes associated with Saturn and Pluto. According to

Tarnas' analysis of this archetypal pairing, periods when Saturn and Pluto are in dynamic alignment, as they were in September 2001, are characterized by an 'atmosphere of gravity and tension', by 'profoundly weighty events of enduring consequence', and by 'suffering under the impact of cataclysmic and oppressive forces of history'.[46] These are often periods of severe hardship, of immense pressure and constriction, of 'historical crisis and contraction', in which many of the darkest, most harrowing of human emotions are accentuated: fear, terror, claustrophobic dread, even a sense of eternal damnation, of being in hell or purgatory, or of facing apocalypse. Such experiences can evoke from the human spirit a response equal to these challenges: moral gravitas and the courage to face the darkness and to overcome fear; Herculean strength and fortitude; relentless exertion and resilience of will to resist overwhelming forces of oppression; the shouldering of the heaviest of burdens or the assumption of some fateful, world-historical calling. Tarnas discovered that the same two planets were in major quadrature alignments at the time of many of the gravest periods of recent history including the start of the First World War, the start of the Second World War, the start of the Cold War in 1946, at the height of the Cold War nuclear arms race in the early to mid 1980s, and then most recently at the time of the 9/11 attacks. By itself, the recognition of the characteristic elements of the Saturn-Pluto archetypal complex in relation to the fateful events of September 11, 2001, is strikingly illuminating, accurately indicating the prevailing mood and themes painfully evident in the tragic experiences of that day and afterwards.

More recently, the square alignment between Uranus and Pluto (2007–2020) has already coincided with many instances of what Tarnas has called the 'unleashing of the forces of nature', such as the BP Gulf oil disaster, the Icelandic volcano eruption, the earthquake-tsunami in Japan and ensuing Fukushima nuclear crisis – events in which elemental forces (Pluto) were suddenly activated and awakened (Uranus).[47] Clearly, then, all three pairs of alignments are directly relevant to the events we are considering here in connection to the ecological crisis and global terrorism.

It should be pointed out, however, that the possible association of the Eris archetype with these events does not contradict or invalidate any of the above analyses involving the planetary archetypes currently recognized in astrology. What we are doing here – to reflect for a moment on the methodology – is examining sets of historical events that were previously analyzed using the other planetary archetypes, such as Saturn-Neptune and Saturn-Pluto, but now taking into account the potential relevance of another archetypal factor. I must stress that it is not that Eris must necessarily be associated with other, quite different sets of events and experiences of its own; rather, incorporating Eris might allow us, I believe, to recognize an additional archetypal dynamic operative within the already-known world events. It might enables us to detect a different hue within the spectrum of archetypal influences present in these events. If the earlier reasoning is correct, it seems to be the case that the discovery of a new planet heralds the recognition of an archetypal principle that was already present and active in human experience but that was only implicit and therefore largely unrecognized within historical events. With the introduction of Eris into our analysis, then, although we are often examining the same sets of events as before, we might now be able to uncover an additional archetypal factor. In the case of Eris, we are alerted, perhaps, to the deeper evolutionary significance of what is unfolding, to the underlying dynamic between civilization and nature, and between all sets of opposites, that drives the evolutionary movement towards unity, wholeness, and the realization of a higher spiritual will or cosmic order. Further details on the significance of Eris for the archetypal astrological analysis of world history are given in Appendix B and Appendix C. The positions of Eris in the Zodiac from 1600 to 2036 are given in Appendix D.

Scientific discoveries

Turning now to science, the first identification of the other modern planets coincided, as we saw in Chapter 2, with highly significant scientific developments and discoveries. The discovery of Uranus broadly coincided with major advances involving electricity in the second half of the eighteenth century and early nineteenth century, such as the invention of Benjamin Franklin's lightning rod around 1750, Alessandro Volta's battery in 1800, and Michael Faraday's electric motor in 1821. The discovery of Neptune broadly coincided with discovery of electromagnetism by Maxwell and Faraday (which provided the foundation for Einstein's relativity theory) and also with the discovery of anaesthetics in medicine. And, as we have seen, the discovery of Pluto coincided with the splitting of the nucleus of an atom and the harnessing of nuclear power. Each of these developments seems to be meaningfully connected to the nature of the relevant planetary archetype: Uranus is typically associated with jolting, stimulating, energizing qualities that are often likened to an electric charge; Neptune, relating to unity, the dissolution of boundaries, the subtle, and intangible, is reflected in the nature of the electromagnetic field, which transcends the limits of the classical Newtonian world picture of separate, indestructible atoms; and Pluto's association with immense power and destructive force is most obviously reflected by nuclear energy.

It was of considerable interest to me, therefore, and it struck me as potentially significant, that the proof of the existence of *dark matter* and the discovery of the existence of *dark energy*, which might turn out to be the major scientific discoveries of our day, occurred around the same time as the sighting and identification of Eris. Cosmologists have made the staggering discovery that the visible universe only accounts for about 4% of the universe's overall composition, and that the remainder is made up of invisible dark matter (22%) and dark energy (74%). Since the 1930s astronomers had known that there was far too much gravity coming from galaxy clusters to be caused by ordinary matter by itself, but until recently, because dark matter does not emit or reflect enough light to be seen, they had only been able to infer its existence through its gravitational effects on ordinary matter. In August 2006, however, astronomers announced that for the first time they had effectively photographed dark matter using a technique known as 'weak gravitational lensing'. For the first time they were able to detect the 'signature' of dark matter from the observation of the collision of two enormous galaxy clusters (made up of hundreds of individual galaxies and trillions of stars, and considered to be the biggest energy event outside of the Big Bang), a collision which has the effect of rupturing dark matter from ordinary matter.

Dark energy, which is utterly invisible too, is considered to be a repulsive form of gravity, opposing normal gravity, causing the acceleration in the rate of expansion of the universe. It was first postulated (and then later rejected) by Albert Einstein in the form of his hypothesis of the *cosmological constant,* which proposes that there is a density and pressure associated with what we considered to be empty space. It is now believed that within every galaxy dark matter and dark energy are evenly distributed, holding each other in balance in a kind of compensatory relationship. Considering the universe as a whole, however, since the early phases of the universe's existence there has been a progressive increase in the amount of dark energy relative to dark matter, as if this progressively changing relationship were in some way connected to the evolutionary unfolding of the universe.

More recently, in 2010 and 2011, continuing the study of the collision of immense galaxy structures, scientists focused in particular on one structure called, of all things, the Pandora cluster, which is the product of four separate galaxy clusters that have collided with each other over the last 350 million years. With the Hubble telescope, the research team was able, using gravitational lensing, to observe the effects of dark matter just after the separate galaxy structures crashed into each other.[48] According to an article by Jason Palmer of the BBC, studying the Pandora cluster has given a picture of:

> ... an extraordinarily rare collision – and [provided] a rare opportunity to learn more about dark matter. It remains enigmatic not least because it interacts very little – if at all – with normal matter, so the dark matter of the Pandora cluster has careered through the crash scene, emerging on the other side. The galaxies and hot gas have lagged behind somewhat, and Dr Massey [Dr Richard Massey of the Royal Observatory, Edinburgh] said that leaves huge swathes of dark matter exposed to further study.[49]

That it should be a galaxy cluster called Pandora yielding secrets of dark matter puts us in mind of the myth of Pandora's Box, and its connection to the goddess Eris. Whether or not this association is to be taken as meaningfully connected to our topic, the conditions in which dark matter can be detected – collisions and crashes of matter and dark matter, processes of repeated separation and coming back together – are consistent with some of the proposed archetypal meanings of Eris (discord, irreconcilable conflicts, compensatory reactions, fragmentation and movements towards unity, and so forth), perhaps suggesting a symbolic correlation between dark matter and the discovery of Eris.

In our earlier reflections on the possible relationship of the archetypal Pluto and Eris, we noted that Eris is most likely a deeper principle than Pluto, representing a dimension of experience that can only be consciously realized after the 'descent' into the Plutonic 'underworld' of the psyche. This seems to be borne out by the association of Eris with dark matter. A BBC website feature called 'Science in the Underworld' explains that in order to detect dark matter it is necessary for scientists to descend into some of the world's deepest underground mines where the density of the rock screens out the cosmic radiation that would obscure the detection of the neutralinos (or 'weakly interacting massive particles' – WIMPs) that are believed to make up dark matter.[50] It is only by going deep underground that scientists can directly detect dark matter, a symbolic reflection, perhaps, of the archetypal relationship between Pluto and Eris. For, as we have seen, it is only by descending into the Plutonic depths of the underworld in the psyche that, during individuation, one can discover and bring to consciousness the archetypal dynamics of the Eris principle, as we have defined it here.

If these associations of Eris with dark energy and dark matter are valid, it might be, then, as suggested in the dream of the dark goddess, and aptly symbolized by the astronomical 'demotion' of Pluto to the status of dwarf planet, that in its magnitude, its pervasiveness, and in its capacity to drive the expansion of the universe, the principle associated with the archetypal Eris is indeed deeper and more fundamental than the principle associated with Pluto.

6. Eris and Astrological Theory

In astrology, the discoveries of Uranus, Neptune, and Pluto in the modern era demanded an unexpected revision of established astrological theory to incorporate the newly discovered planets. In particular, astrologers needed to understand how the new planets and their archetypal meanings were connected to the twelve signs and houses of the horoscope.

Rulership in astrology

Whereas the planets represent a set of foundational dynamisms and powers in human experience,, the signs of the Zodiac represent particular qualities that inflect the expression of the planetary archetypes. In natal astrology (the study of birth charts), the nature of the sign in which a planet is positioned qualifies, colours, and conditions the expression of the corresponding planetary archetype in the individual's personality and biographical experiences. For example, the Mars archetype is associated with decisive action, the pursuit of goals, exertion, assertion, and aggression. In itself, it is pointed, linear, and very direct in character. If Mars is positioned in Pisces, however, the energy of the Mars archetype will be markedly altered by the qualities associated with that sign, such as sensitivity, compassion, dreaminess, and a lack of focus. Thus Mars in Pisces tends to correlate with a diffused, scattered, escapist, and more sensitively expressed mode of being than other Mars placements, with the capacity for decisive action and self-assertion compromised by empathy and compassion for others. Mars in Libra, to give another example, tends to be diplomatic, to want to keep the peace, and to keep everything fair and equal, or graceful and harmonious, in keeping with the nature of the sign Libra. Expressions of anger might be motivated more by a perceived lack of decorum or good manners, than by a need to get what one wants. The sign position of a planet, then, can markedly affect the way the corresponding planetary archetype is expressed.

Each sign is said to be 'ruled' by one, or in certain cases two, of the ten planets, based upon the inherent similarity of the archetypal characteristics and themes associated with the ruling planet and the qualities associated with the sign it rules. Each sign belongs to one of four *elements* (Fire, Earth, Air, Water) and is also grouped into one of three *triplicities* (Cardinal, Fixed, Mutable). The *Cardinal* signs (Aries, Cancer, Libra, and Capricorn) have the qualities of initiation, leadership, action, and expending energy; the *Fixed* signs (Taurus, Leo, Scorpio, Aquarius) have the qualities of fixity, sustained form, maintaining energy; and the *Mutable* signs (Gemini, Virgo, Sagittarius, Pisces) have the qualities of adaptation, changeability, and the dispersal of energy into new forms.

In addition to the division into signs, astrological charts are divided into twelve sectors known as *houses*. Each house is related to a set of meanings associated with particular areas of life experience. For example, the first house is to do with the body in action, the appearance, the natural persona, one's instinctive approach to life, and the process of coming to self-awareness; the fourth house is to do with the home, the family, the past, roots and origins, foundations, domestic life, and the private sphere of personal experience; the ninth house is connected to travel, study, philosophies, ideologies, and world views, and any activities that involve reaching out to the wider world, such as writing and publishing. Like the signs, each house is strongly associated with one planet which is said to rule the house, each house belongs to an element grouping, and each corresponds to a sign in the Zodiac. For example, Aries, the first sign of the Zodiac, has a strong correlation with the first house, and both these relate to the planet Mars. Sagittarius, the ninth sign of the Zodiac is associated with the ninth house, and both are ruled by Jupiter. The houses are also divided into three modalities: Angular, Succedent, and Cadent, corresponding to the divisions of the signs into Cardinal, Fixed, and Mutable groups. In the most general terms, *Angular* houses (1, 4, 7, and 10) relate to initiatory action in the world; *Succedent* houses (2, 5, 8, and 11) relate to experiences that both maintain and react to the initiatory actions; and *Cadent* houses (3, 6, 9, and 12) are related to understanding and adaptation.

Sign	House	Ruling Planet	Element	Triplicity/ Modality
Aries	1	Mars	Fire	Cardinal/ Angular
Taurus	2	Venus	Earth	Fixed/ Succedent
Gemini	3	Mercury	Air	Mutable/ Cadent
Cancer	4	Moon	Water	Cardinal/ Angular
Leo	5	Sun	Fire	Fixed/ Succedent
Virgo	6	Mercury	Earth	Mutable/ Cadent
Libra	7	Venus	Air	Cardinal/ Angular
Scorpio	8	Pluto	Water	Fixed/ Succedent
Sagittarius	9	Jupiter	Fire	Mutable/ Cadent
Capricorn	10	Saturn	Earth	Cardinal/ Angular
Aquarius	11	Uranus	Air	Fixed/ Succedent
Pisces	12	Neptune	Water	Mutable/ Cadent

Figure 11. Rulership Cross-Reference Table.

Prior to the discovery of Uranus in 1781, each of the seven ancient planets was connected with one or more signs of the Zodiac; each planet *ruled* one or more of the signs and houses (see Figures 11 and 12). Thus Saturn, which pertains to the archetypal principle associated with time, endings, mortality, discipline, worldly concerns, structure, tradition, the establishment, and a wide range of other related themes, is said to rule the sign Capricorn, which is connected with qualities such as hard work, ambition, practicality, reserve, perseverance, conservatism, realism, self-control, and such like. The characteristics of the sign, then, are generally reflective of the nature of the Saturn archetypal principle. Similarly, the Sun, associated with the archetypal principle of self-expression, life energy, vitality, centring, identity, and consciousness is deemed to rule the sign Leo, which is typically related to qualities and traits such as a self-expressive character, warmth, self-centredness, enthusiasm, personal pride, and a dramatically expressive mode of being. Both the planet and the sign are connected to the urge to be, to express oneself, to shine, and radiate warmth and creative energy.

Ruling Planet	Signs Ruled	Houses Ruled
Sun	Leo	5th
Moon	Cancer	4th
Mercury	Gemini, Virgo	3rd, 6th
Venus	Taurus, Libra	2nd, 7th
Mars	Aries, Scorpio	1st, 8th
Jupiter	Sagittarius, Pisces	9th, 12th
Saturn	Capricorn, Aquarius	10th, 11th

Figure 12. Ancient Rulership Table.

Incorporating the outer planets

After the discovery of Uranus, Neptune, and Pluto, for the system of rulership to continue to make sense, these newly discovered planets had to be assigned rulership of a sign and house. Rather than undermine astrological theory, however, as might have been expected, in many ways the inclusion of the new planets made the system more coherent (see Figure 13). Uranus, it was discovered, had much in common with the sign Aquarius (and the eleventh house), pertaining to qualities and themes such as individualism, humanitarianism, social awareness, intellectualism, science, technology, a future orientation, and eccentricity. Uranus was, in many respects, a better match to the sign Aquarius and the eleventh house than the ancient ruler Saturn. For many astrologers thereafter, Uranus was thus considered to be the true ruler of Aquarius. For others, rulership was to be shared between Saturn (the ancient ruler) and Uranus (the modern ruler).

A similar process took place in the case of the planetary archetype associated with Neptune, which was shown to share many of the qualities inherent to the sign Pisces and the twelfth house, and appeared to be a better match to these than the ancient ruler Jupiter. So too Pluto was discovered to be similar in its archetypal meaning and attributes to the qualities and characteristics of the sign Scorpio and the eighth house, formerly ruled by Mars. In summary, then, while the conception of rulership is not without its problems and ambiguities, it is generally accepted that the new system is more accurate than the traditional model, better reflecting the inherent similarities between the planetary archetypes and the signs and houses they are connected with.

Ruling Planet	Signs Ruled	Houses Ruled
Sun	Leo	5th
Moon	Cancer	4th
Mercury	Gemini, Virgo	3rd, 6th
Venus	Taurus, Libra	2nd, 7th
Mars	Aries	1st
Jupiter	Sagittarius	9th
Saturn	Capricorn	10th
Uranus	Aquarius	11th
Neptune	Pisces	12th
Pluto	Scorpio	8th

Figure 13. Revised System of Rulership Incorporating Uranus, Neptune, and Pluto.

Rulership and Eris

The revisions resulting from the discoveries of Uranus, Neptune, and Pluto have left only two remaining planets allocated rulership of more than one sign and house: Mercury, which is said to rule both Gemini and Virgo, and the third and the sixth houses; and Venus, which is said to rule both Taurus and Libra, and the second and the seventh houses. It would make sense, therefore, if Eris is to be designated ruler of a particular sign and house, that it would be one of these two sign-house combinations. Of course, following the spate of recent discoveries of dwarf planets, plutoids, and other significant asteroids, there are now too many planet-like entities in the solar system for each sign and house to be ruled by only one planetary body. As I noted earlier, however, of all the newly discovered planetary bodies, Eris does seem to be of particular importance, both because of its relatively large size, the prominence of its discovery, and because its discovery directly led to the reformulation of planetary categories and definitions in the solar system. How the other planet-like objects might be incorporated into this system is a matter for future discussion.[1] But the discovery of the new planets will certainly entail a radical evolution in astrology, dramatically increasing the number of archetypal variables to be considered.

Of the remaining signs (Taurus, Libra, Gemini, and Virgo) that currently share a planetary ruler, Venus, as we will see, seems more appropriately connected with the Earth sign Taurus and the second house (a sign-house pairing that is typically associated with sensual pleasure, the natural world, money, and resources) than it does with Libra. Mercury, of the two signs and houses with which it is associated, seems more directly compatible with Gemini and the third house, as these relate to themes such as communication, thinking, questioning, and learning. The planet-like asteroid Chiron, discovered in 1977, and connected to the archetype of the 'wounded healer', has been associated by some astrologers with Virgo and the sixth house, which have to do with themes such as health, healing, work, service, an attunement to the material world, craftsmanship, and practical or technical skill.[2] This leaves the sign Libra and the seventh house, which, I believe, are both clearly compatible with themes we have associated with Eris.

The seventh house is associated with the area of life concerned with relationships and partnerships, and, more abstractly, this house is very much associated with 'the other' in its many forms, with the experience of the 'not-self', and with the growth of self-awareness through the encounter with the other in relationship – themes we speculatively connected to Eris, above. The sign Libra is associated with traits such as a strong sense of personal fairness, striving for harmony and balance, justice, vacillation between competing positions, and a pronounced focus on relationships. Because of this striving for harmony and fairness, people with a strong emphasis on Libra tend to experience an often acute sense of disharmony and discord, which they are therefore prompted to seek to resolve and reconcile. Liz Greene, one of the foremost astrologers of the last thirty years, has made a similar point. She also calls into question the sign Libra's association with themes connected to Venus, such as romantic love:

> I have never felt that Libra was concerned, as some popular
> descriptions would tell us, with romantic love, flowers

and candlelight, except as an abstract concern with the
appropriate rituals of courtship according to an ideal
conception. Romantic 'feeling' is not a property of Libra.
The sign is much more connected with questions of
ethics and morality, judgment and apportionment ... the
perfectly balanced scales of judgment.[3]

She adds, however, that 'imbalance and extremes and the violation of
the law are necessary happenings from which Libra does not readily
escape', which is, of course, compatible with themes we have discussed
in relation to Eris.[4] Identifying mythic figures closely aligned with
Libra, Greene connects the sign with the Greek characters the Erinyes
(or Eumenides) and Dike (justice), associated with morality and good
judgment. Yet the mythic Eris, too, our reasoning suggests, seems to
embody a number of qualities compatible with Libran traits. As with
each sign of the Zodiac, it is to be expected that a number of different
mythic figures will personify different attributes and qualities of the
sign.

Further supporting our speculations regarding Eris and its possible
rulership of Libra, Greene connects the sign to the myth of the Judgment
of Paris. A major challenge associated with Libra, Greene believes, is
the necessity for exercising good judgment when confronted with an
inescapable decision thrust upon one against one's wishes – as in the Paris
myth. Every such decision, however, involves a potential violation:

> The choosing of one thing over another, which life seems to
> force upon Libra, not only contradicts the sign's innate desire
> for having everything in proportion rather than one thing
> at the expense of another. Such a judgment also involves
> psychological consequences, for any decision of an ethical
> kind made by the ego means the exclusion or repression
> of some other content of the psyche, which produces
> enormous ambivalence and sometimes great suffering.[5]

These themes are, of course, very similar to those we have attributed
to the archetypal Eris: violation of certain aspects of the psyche, the
separation of ego-consciousness from the state of natural unconscious
wholeness, the conflict between two or more seemingly irreconcilable
positions, and the impulse to seek harmony from discord. Further

addressing the relationship between harmony and discord, Greene adds:

> It would seem that the development of Libra encompasses a curious paradox: that the sign is in love with the orderly laws of life and places great faith in their fairness, yet is perpetually confronted by the disorderly and immoral aspects of life, which fragment and divide Libra's cherished unity ... [W]hile the sign cannot bear division or disharmony in the universe, something within the Libran himself forever drives him to divide himself.[6]

Given this paradox, 'it is not surprising,' she observes, 'that Libra perpetually complains about the unfairness of life'.[7] But this unfairness, she concludes, might be but the foreground to a background order of greater justice.

In summary, then, based on the provisional meanings attributed to Eris in this essay, the sign Libra and the seventh house appear to be strong candidates to fall under the rulership of the Eris archetype based on similarities of meaning. Indeed, Eris' possible association with the principle of balance and justice is also aptly suggested by the symbol of the sign Libra: the scales.[8]

Ruling Planet	Signs Ruled	Houses Ruled
Sun	Leo	5th
Moon	Cancer	4th
Mercury	Gemini	3rd
Venus	Taurus	2nd
Mars	Aries	1st
Jupiter	Sagittarius	9th
Saturn	Capricorn	10th
Chiron	Virgo	6th
Uranus	Aquarius	11th
Neptune	Pisces	12th
Pluto	Scorpio	8th
Eris	Libra	7th

Figure 14. Possible System of Rulership Incorporating Eris and Chiron

A planetary glyph for Eris

If Eris is to be incorporated into astrological theory, it will need to be assigned its own symbol or planetary *glyph*, which is used to represent the planet in astrological charts and ephemeris tables. To date, no glyph for Eris has been agreed, although several have been proposed. Here I would like to present another alternative, based on the various insights into the archetypal meaning of Eris that we have considered in this essay. Convention dictates that astrological glyphs are in line format only, and I will adhere to that guideline here.

It should be pointed out that the glyphs of the other modern planets do not necessarily reflect the meaning of the corresponding planetary archetype, but have been based on other considerations, such as the name of the planet's discoverer. For example, the glyph for Uranus (see Figure 2, page 16) incorporates the letter H, first letter of the name of its discoverer, Herschel, just as one common symbol for Pluto (♇) not only draws on the first two letters of the planet's name, but on the initials of Percival Lowell, the astronomer whose efforts eventually led to the discovery of Pluto by Clyde Tombaugh in 1930. In the case of Neptune's glyph, a trident, this was chosen solely with the mythic Greco-Roman god in mind. With regard to an Eris glyph, however, it would be helpful, I think, if the glyph were to symbolize the archetypal qualities we have here associated with the corresponding planetary archetype. Contemplation of the glyph might thus evoke some of the associated astrological themes.

Considering the archetypal dynamics we have explored here in connection with Eris, I was struck by how aptly the following drawing (Figure 15) by the German mystic Jakob Böhme (1575– 1624) captures a number of the essential elements and the deepest significance of these dynamics. In this drawing, Böhme, who was an important influence on figures such as Blake and Jung, portrays the divine in the form of two semi-circles in opposition to each other. The rotated hemispheres meet at the intersection of a crucifix symbol, suggesting the tension of opposites running through the realization of wholeness. Böhme's drawing is also a symbol of the juxtaposition of the light and dark aspects of divine, incongruently related, and centred on the human heart.

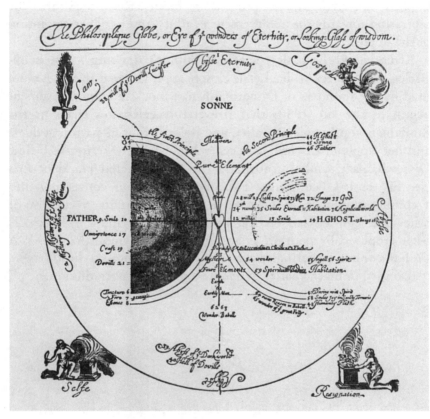

Figure 15. The Philosophical Globe, *Jakob Böhme.*

According to Jung, this image portrays the irreconcilable nature of the opposites that are nevertheless brought into a unity within the mandala drawing. 'Their periphery,' Jung notes, 'contains a bright and dark hemisphere turning their backs to one another. They represent un-united opposites, which presumably should be bound together by the heart standing between them.'[9] The light Holy Ghost opposes the dark Father, a polarity which is 'crossed by the paired opposites "Sonne" and "Earthly Man".'[10] The picture, in Jung's view, 'aptly expresses the insoluble moral conflict underlying the Christian view of the world'.[11] More generally, the drawing portrays the tension of opposites – good and evil, light and dark – which generates the whole drama of existence. This deep-rooted polarity, or discord as we have called it here, can be imagined as the primary archetypal impulse underlying the life process, an impulse that draws forth the aspiration towards the achievement of unity and a higher resolution of the opposites.

Remarkably, the basic pattern of Böhme's drawing – two semi-circles turned back to back – is closely approximated in the symbol used by the founders of Discordianism, a mock religious movement begun in the late 1950s that irreverently celebrates chaos as the fundamental principle of reality, and holds Eris as its patron goddess and her 'apple of discord' as a central motif in its 'scripture'.[12] The Discordianism symbol, known as the Five-Fingered Hand of Eris (see Figure 16), comprises 'two opposing arrows converging into a common point' that may be drawn horizontally, vertically, or otherwise.[13] A stylized vertical version of this symbol has actually been proposed by petition as an astronomical symbol for Eris.[14] A similar symbol, featuring two arrows positioned head to head (➤◄), is also used in mathematics to denote a proof by contradiction.[15]

Figure 16. The Five-Fingered Hand of Eris, Symbol of Discordianism.

Modifying slightly the vertical Discordianism symbol, and with Böhme's drawing in mind, I would like to propose (see Figure 17) the following glyph for Eris:

Figure 17. A Glyph for Eris.

This glyph is intended to point to the antagonism and inherent discord within the overall unity of life. The opposing hemispheres reflect the division of the unconscious wholeness of existence, a division that is manifest in the relationship between civilization and nature, consciousness and the unconscious. The semi-circles, turned back to back, also suggest the challenge of recognizing the other – our own unrecognized opposite half, as it were – which shows us life's opposite face, and with which we are always in contact, even as we face the other way. The contact between the semi-circles – the fact they press against each other – suggests, further, the creative friction and strife of irreconcilable opposites, and their unavoidable collision that underlies all change and that is the motive force behind the upward thrust of human evolution towards ever greater consciousness (suggested by the vertical axis of the crucifix). The crucifix symbol, finally, here stands for the transformation of the structure of the human psyche that results from the confrontation between opposites, and the possibility of holding the opposites in dynamic balance.

Conclusion

On this note we draw our survey to a close. In writing this essay, it has been my intention to attempt to identify the possible archetypal meaning of Eris as a basis for the long process of research that must now be undertaken. I felt that it might help future research studies exploring astrological world transits, personal transits, and birth charts to have some idea of what Eris might signify, as a starting point. Needless to say, the hypothesis outlined here must be viewed as entirely provisional. The various clues and indications emerging from all of the above principles as to the archetypal significance of the discovery of Eris are only based on informed conjecture, and it remains to be seen whether the passing of time will confirm or disprove this hypothesis. Nonetheless, the clear parallels between the various perspectives and principles we have discussed in delineating the archetypal meaning of Eris are sufficiently suggestive and mutually concordant, it seems to me, to warrant serious consideration.

In summary, the heightened expression and increasing prominence in our time of the cluster of archetypal themes we have associated with the Eris archetype appears to present a number of challenges at multiple levels of our experience:

Ecological: coming to terms with and managing the destructive and violent reaction of nature/the Earth against its violation by human civilization; striving to bring human civilization into a more harmonious relationship with the natural order.

Socio-cultural and political: learning to relate to the 'other'; dealing with the problem of excluded groups; balancing tensions between fragmentation and unity; the challenge of finding a path to greater planetary unity and harmony;

integrating the reactions of repressed feminine emotional-instinctual power against the old dominant patriarchal structures of modern civilization; reactions against the extraverted superficiality of modern culture.

Technological and scientific: bringing science and technology into harmony with the ways of nature rather than using technology as an instrument to fulfil human ends; working with nature rather than violating it.

Psychological: coming to terms with the opposites inherent in the human psyche; undergoing the psychological transformation that results from the ego's encounter with the unconscious; addressing the psychological division between ego-consciousness and its instinctual foundations; moving through the opposites to greater unity; using the tension, challenge, stress, and discord in one's life as a spur to excellence and to the realization of wholeness.

Cosmological: the exploration of the hidden universe of dark matter and dark energy; the recognition of the complementary forces holding our universe in dynamic equilibrium, and their corollaries in the human psyche.

Evolutionary: coming to terms with a significant, even epochal turning point in the evolution of the human species and the planet; understanding and working through the 'turning around' in the evolution of human consciousness and culture.

Spiritual: growing beyond egocentricity with its often narrow sense of personal fairness and justice, to embrace a divine or cosmic justice; cultivating the realization that all opposites reside in a deeper unity; embracing all of the above as a the expression of a spiritual unity, a divine ground; aligning personal will with divine will.

While the words *discord* and *strife* usually carry wholly negative connotations, by considering the mythic and philosophical origins of these terms I have attempted to direct our attention to the deeper principles they represent. In so doing, I have also endeavoured to draw out the constructive significance of discord for evolution and spiritual development. For it is only through an affirmation of the discord inherent in life that one can attain the spiritual posture of equanimity and balance in the face of changing life circumstances. In this sense, discord serves an impetus to psychospiritual transformation. The Kingdom of Heaven, we read in *The Gospel of Thomas*, is a 'movement with a repose'.[1] Heaven, that is to say, is the realization of a psychological condition of dynamic harmony, rather than a static endpoint devoid of tension. It represents, in psychological terms, a condition in which the often antagonistic instincts and impulses in the human psyche become organized around a deeper centre and subsumed within a larger wholeness. Discord, arising from the separation of the ego and the unconscious, subject and object, and spirit and nature, might be thought of as a vehicle for the 'making whole' of the human personality.

At the global level, too, we have the opportunity to use the discord we experience as a stimulus to build a planetary civilization that can meet the immense challenges ahead, not least those posed by climate change, international conflict, terrorism, economic hardship, overpopulation, ecological devastation, and shortages of essential resources. The creative resolution of the challenges of the current world situation will depend, to a large extent, on how we react and respond to planetary discord from here on. It will depend, too, on how we deal with the 'other' and the problem of the opposites, both within ourselves, and in the world at large. Such challenges, if this thesis is correct, are integral to the archetypal themes and experiences associated with the new planet Eris.

While the themes and arguments set out above might give us cause for concern about our future path (as we move inexorably into the deepening ecological crisis and live with the undiminished threat of global terrorism), our prospects for skilfully navigating the challenges ahead can only improve with a greater awareness of the archetypal and evolutionary dynamics at work. To this end, whether or not one accepts the astrological correlation with the planet Eris, the exercise of reflecting upon the archetypal undercurrents of the world situation remains of great value. It might not be possible to avert some of the painful changes ahead, but we can at least try to remain mindful of an underlying *telos* at work and live with an eye on the deeper meaning of whatever comes to pass. 'Meaning,' as Jung once said, 'makes a great many things endurable – perhaps everything.'[2]

APPENDIX A

Possible Meanings of the Eris Archetype: A Summary

* Pairs of opposites that underlie all change in the world, and between which life energy moves and is held in dynamic tension.
* Compensatory responses; the reaction of the unconscious; reactions arising from polarity; the regulatory, counterbalancing forces of nature.
* Strife and discord relating to the principle of deeply-rooted evolutionary change that underlies the separation between civilization and nature, between ego-consciousness and its instinctual foundations.
* The conflict and discord inherent in the nature of things that stimulates the evolutionary movement towards the realization of the global unity of the Earth as a living planet.
* The experience of violation and perceived injustice as a result of evolutionary change.
* The backlash of nature and of human nature against violation and injustice, sometimes resulting in indiscriminate destruction and the large-scale loss of life.
* The experience of resentment arising as a response to the discord wrought by evolutionary change.
* The expression of indiscriminate rage and violent lashing-out fuelled by the resentment of groups that feel violated or excluded.
* The closing of the division between civilization and nature, sometimes in the form of a collapse of civilization or catastrophes that bring human collective consciousness

back into alignment with nature, back in touch with reality,
putting to an end more superficial modes of living.
* Evolutionary turning points and great reversals.
* The movement through strife and discord towards harmony,
through bringing individual human lives into alignment with
the will of the One – the cosmic or divine will.
* Fragmentation that precedes the formation of higher-level
unities.
* The 'dark spirit' in matter and nature, often conceived in
myth as a 'feminine' spiritual principle, as a compensatory
force to the transcendent divine; the polarity between these
expressions of the divine.
* Dark matter and/or dark energy in the cosmos, existing in
complementary relationship, counterbalancing each other;
dark energy opposing the force of gravity.
* The principle of cosmic justice, and the strife between the
human order and this justice; the personal will versus the
cosmic or divine will; the challenge of overcoming the sense
of personal injustice and fairness to embrace the will of the
whole.
* The experience of an irreconcilable tension of opposites during
individuation that stimulates psychological transformation;
the juxtaposition or coming together of conflicting opposites;
passing through the opposites.
* Discord and disputes requiring evolutionary advances, greater
global unity, or psycho-spiritual growth to resolve; 'conflicts
of duties' that force a regression of libido in the psyche – that
is, irreconcilable conflicts between forces of equal power
in the psyche that demand the shift to a higher level of
consciousness to resolve such that the problem is transferred
to a higher level.
* The need for a differentiation of the emotional sphere to come
to terms with the deep-rooted discord, rather than falling
victim to it through its external forms of expression.
* The path and transition to a new level of consciousness; the
principle that impels one to overcome all discord, to subsume
all antagonistic reactions within a higher unity.
* Competition as a spur to excellence and wholeness.

AUTHOR'S NOTE

Upon completion of this work, I discovered that certain of these possible meanings of Eris have also been proposed in a number of internet articles on the subject. For example, Roy McKinnon suggests the Eris principle might have to do with the 'conflict between desires of personality and pursuit of the inner call: persecution of the spiritually aware person who stands alone courageously acting according to conscience' alluding perhaps to the discrepancy between personal will and divine will, human justice and cosmic justice, we have explored here (see http://zanestein.com/keywords.html#Eris). He also lists, as another complex of themes associated with Eris, 'striving to achieve one's goals and refusing to capitulate to the pressure of unjust treatment and discord from abusive authority: with great inner conviction and single-mindedness of purpose following a path of high attainment and enlightenment notwithstanding adversity—[or] alternatively refusing the call to transform with mediocrity and underachievement as consequences.' Again, this description draws together themes covered here, such as affirming and rising above personal injustice, and using adversity as a spur to excellence.

Francesco Schiavinotto's list of possible Eris keywords also overlaps, to a degree, with some of the themes considered in this book (again, see http://zanestein.com/ keywords.html#Eris). Among the keywords he proposes are the following: 'To be opposite; Aversion for the Extraneous; To obstacle [sic] the Alien; False Adaption between Different Matters; Reciprocal Aversion; Feeling to be Invaded . . . Ecological Incompatibility; Underground Instability.' The feelings of invasion and 'ecological incompatibility' seem to relate to the violation of nature theme and the discord between human civilization and the natural world. 'Aversion for the extraneous' perhaps relates to what we have called the problem of the other and dealing with excluded groups. Schiavinotto also alludes to the tension of opposites through terms such as *aversion* and *incompatibility*, and the 'underground instability' can perhaps be connected to the suggestion that Eris is a deeper dynamic of the Pluto archetype, relating to the movement of life between opposites that gives rise to the power associated with Pluto.

Zane Stein suggests that Eris might relate to the 'loss of innocence' that is connected with the 'entrance into adulthood' perhaps alluding to the theme of violation, as I have defined it here. He also connects Eris to an 'internal split causing [the] longing to be whole', which is a close approximation to our explorations of the separation of ego-consciousness from the unconscious and the psychological encounter with the opposites on the journey to wholeness. Although McKinnon, Schiavinotto, and Stein suggest other possible meanings not as consistent with the themes we have explored here, to my mind the overlap between the ideas explored in this book and their own suggestions is noteworthy and significant.

APPENDIX B

Eris World Transits I: Saturn-Eris Alignments

The events surrounding the outbreak of the First World War, understood in terms of the relationships between the Neptune, Pluto, and Eris principles, seem to be coherently reflected in world transits of the time. World transit analysis is based on a consideration of the geometric alignment – the specific angle of relationship – formed between the different planets on their orbits. The changing pattern of planetary relationships is studied in astrology to understand the changing relationships between the corresponding archetypal principles. To know how we are related to the planets at a moment in time gives us insight into how we are related to the different archetypal principles these planets represent. The study of the correspondence between these changing planetary alignments and the patterns of world history or individual biography is known as *transit astrology*. The set of significant angular relationships (or 'aspects') between the planets at any given time are called *world transits*.

During the first Hague Convention of 1899 (which served as basis for the later establishment of the League of Nations), Eris was in a square (90-degree alignment) to the Neptune-Pluto conjunction of the time, and also square to Saturn, the three groups of planets forming what is known as a T-square. By 1907, the time of the second Hague Convention, Eris remained in the square alignment to Neptune-Pluto and was also by this time in a conjunction with Saturn. Seven years later, at the outbreak of the First World War in July 1914, Eris was in another exact square alignment with Saturn and it remained in the same square with Pluto. In archetypal terms, the role of Saturn in these events and themes, which were so prominent during the years leading to the outbreak of the war, relates primarily to the protection of national interests, attempts to establish or reinforce boundaries, and to preserve the dominion of the old colonial empires (thus resisting the impulse towards fragmentation and the formation of larger unities that we have associated with Eris). These events reflect Saturn's association with self-protection, boundaries, fear, and defensiveness. The conventions also sought to regulate military conflict by establishing laws overseeing the settlements of disputes and establishing the rules of war – themes we might expect to see given the meanings prospectively associated with Eris here (see Figure 18).

Alignment	Dates	Events
Conjunction	May 1966–April 1969	Vietnam War escalates Peace Movement and Anti-War protests Ecology Movement begins
Square	September 1973–July 1975	ITT Corporation bombed by terrorists in New York Endangered Species Act passed in US First UN International Women's Day Vietnam War ends
Opposition	September 1980–August 1983	AIDS identified First American test-tube baby born First UN International Day of Peace Environmental Justice Movement begins (autumn 1982)
Square	January 1989–January 1991	Earth Day protests April 1990 Bangladesh cyclone Armenian earthquake EU nations agree to ban CFCs Exxon Valdez Oil spill followed by Oil Pollution Act Langkawi Declaration by Commonwealth for environmental sustainability Ecole Polytechnique massacre in Montreal leading to National Day of Remembrance and Action on Violence Against Women

Conjunction	May 1996– July 1998	Genetically modified food controversy Cloning of first sheep Federal funding of human cloning blocked
Square	August 2003– July 2005	Kyoto Protocol on climate change comes into force Indian Ocean tsunami Terrorist bombings in London
Opposition	September 2010– September 2013 (events listed as of July 2011)	Japanese earthquake/tsunami Fukushima nuclear disaster Killing of Osama bin Laden Norway twin terror attacks

Figure 18. Some Possible Correlations with Recent Saturn-Eris Alignments.
15 degree orbs used for conjunctions and oppositions; 10 degree orbs used for squares.

During the period of the formation of the United Nations in October 1945, Neptune and Venus were in opposition alignments with Eris; and during the formation of NATO in April 1945, Eris was in the same opposition alignment with Neptune, and was then also in a conjunction with the Sun, Venus, and Mars. The Neptune-Eris opposition of this period perhaps correlates with the moment when the irreconcilable conflict of the Second World War collapsed into total surrender (a theme associated with Neptune) in the wake of the immense horror of the atomic bombing of Hiroshima and Nagasaki, and, thereafter, the idealistic aspiration (associated with Neptune) on the part of many countries to make and secure a lasting peace.

In a preliminary survey of world transits involving Eris, I noticed that recent Saturn-Eris hard-aspect alignments (conjunction, square, opposition) tend to correlate with attempts to prevent the violation of nature, with Saturn, the principle of restraint, limitation, negation, and so forth, seemingly negating or limiting themes we have attributed to Eris. It was during the Saturn-Eris conjunction of the late 1960s that negative judgments about what constitutes 'progress' decisively entered public consciousness with the increasing awareness of damage caused to the natural world by the development of modern Western civilization. The Endangered Species Act was passed in the US during the following Saturn-Eris square of September 1973 to July 1975. During the subsequent square of January 1989 to January 1991, the Earth Day protests took place (in April 1990), the same alignment during which EU nations agreed to ban CFCs in an attempt to protect the ozone layer. This transit also coincided with the Exxon Valdez Oil Spill which led to the passing of the Oil Pollution Act, and with the Langkawi Declaration by the Commonwealth for environmental sustainability. During the conjunction of 1996–1998, the first cloning of sheep took place (violation of nature theme) and federal funding of human cloning was blocked (the negation and control of the violation of nature). It was at this time, too, that the problematic consequences of genetically modified food impressed themselves on the public consciousness, especially in Europe. A few years later, the Kyoto Protocol on climate change came into force during the square of July 2004–July 2005 as a legislative measure to control and regulate human impact on the natural environment. Finally, during the Saturn-Eris opposition in 2011 the Fukushima earthquake-tsunami-nuclear disaster took place, bringing it home to many people around the world how fragile supposedly secure nuclear facilities are, further alerting us to the threat posed by failed nuclear power stations to the natural environment and to human life, causing nations to reconsider future nuclear policies.

As noted, to understand the specific character of each of these events, one needs to take into account the other major world transits at the time. For example, the theme of the control and domination of nature is associated with major Saturn-Pluto alignments (see Tarnas, *Cosmos and Psyche*, pp.253–56). The impulse to impose human will on nature, to harness nature's titanic elemental forces, or to protect ourselves from natural forces each relate to this combination. Equally, the events of the 1988–1991 Saturn-Eris alignment reflect themes associated with the Saturn-Uranus-Neptune conjunction of that period, such as pollution and poisoning (Saturn-Neptune) and accidents (Saturn-Uranus). It seems to be the case, our thesis suggests, that Eris relates to the deeper archetypal dynamics underlying these other, more specific archetypal factors.

Also evident during these same Saturn-Eris alignments was the theme of resistance to discord. Perhaps the clearest example of this was during the Vietnam War, which reached a critical point in the late 1960s, as Saturn and Eris came close to an exact conjunction.

On January 31, 1968, the North Vietnamese army and Vietcong launched the surprise 'Tet Offensive' against US military on several fronts, which, although repelled by US forces, had the effect of decisively shifting American public opinion against the war, largely due to the broadcast of television images showing the scale of the casualties. Indeed, this is widely seen as the major turning point of the war. In addition to the immense discord of the conflict itself, we can see here again several themes we have explored in relation to Eris, including major turning points, the growing resentment against the war (directed at those in authority – associated with Saturn), the desire to end the conflict, the irreconcilable feud between communist and capitalist ideologies bringing the two major world superpowers into direct confrontation, and the all-too-apparent gross injustice (Eris) of the war. In the aspirations of the Peace Movement, Anti-War Movement, and the hippy counterculture, we can see examples of resistance to discord and attempts to negate strife wrought by economic 'progress' and the inexorable march forward of human civilization. Similar themes are evident in subsequent hard-aspect alignments between Saturn and Eris. For example, the opposition of September 1980 to October 1983 accompanied the first UN International Day of Peace. And during the square of 1988–1991, the Ecole Polytechnique massacre took place in Montreal, which led to the establishment of the National Day of Remembrance and Action on Violence Against Women – an example here of attempts to prevent further instances of the violation of human nature.

Obviously these few brief examples are far too selective to substantiate any meanings attributed to Eris above, but they do suggest that the meanings we are considering might at least be on the right lines.

APPENDIX C

Eris World Transits II: Jupiter-Eris Alignments

It is perhaps not without significance that many of the natural disasters and terrorist attacks, discussed above, occurred during hard-aspect alignments (conjunction, square, and opposition) between Jupiter and Eris. Jupiter is associated with the archetypal principle of expansion, elevation, enormity of scale, abundance, success, affluence, fruition, public recognition. breadth of scope, and wide cultural vision. It is associated with a dramatic larger-than-life quality, with a positive attitude and generosity of spirit, and with confidence, sometimes over-confidence. The Eris principle – to summarise some of the main themes of our hypothesis – perhaps relates to discord and its connection with evolution, the experience of injustice prompting an awareness of or attempts to express a higher principle of justice, indiscriminate rage, the potentially violent reaction of the excluded fuelled by resentment, the strife associated with the movement towards globalization, the schism between humanity and nature, the retaliatory reaction of nature, the breaking apart of civilization, counterbalancing or regulating reductions in population levels, and to epochal turning points and great reversals. On September 11, 2001, Eris was in a broad square alignment with Jupiter (part of a T-square with Mercury), which became more exact in the days after the attack. At the time of both the Indian Ocean tsunami and Hurricane Katrina, Eris and Jupiter were once again in dynamic angular relationship – the very next quadrature alignment – as Jupiter, in its twelve year orbit, had come into a close opposition with the extremely slow-moving Eris. Whereas in September 2001 Jupiter had been in the sign Cancer, ninety degrees from Eris in Aries, it had now moved on to Libra, the sign directly opposite Aries in the Zodiac. Moving into May of 2008, at the time of the Burma cyclone and the China earthquake, Jupiter and Eris were in a tight, almost exact square alignment. The same square alignment between Eris and Jupiter was still in range at the time of the global financial crisis (September 2008) and it was close to exact when the terrorist attacks in Mumbai took place in November of that year. The previous conjunction of Eris and Jupiter was in 1999 at the time of the Kosovo War in Yugoslavia and the Columbine high school shootings in Colorado. More recently, the March 2011 earthquake, tsunami, and the ensuing Fukushima nuclear crisis in Japan occurred under a Jupiter-Saturn-Eris alignment, with Jupiter in conjunction with Eris in Aries and in opposition to Saturn in Libra. The same alignment was still in range on July 27, 2011, at the time of the twin terror attacks in Norway, when a lone gunman killed sixty-nine people at an island youth camp and eight others in a bomb explosion in Oslo.

From an archetypal perspective, what seems to be common to these occurrences is that they are all particularly notable, international, or large-scale manifestations of events in which certain of the themes we have provisionally attributed to Eris were magnified or took on a decidedly international cast. Indeed, several themes we have attributed to Eris were combined with those associated with Jupiter in a number of ways. The natural disasters obviously represent dramatic expressions of the counterbalancing forces of nature wreaking destruction on the structure of the human-built world. And in the case of 9/11, we can see both resentment and the violent reaction of groups who feel excluded or who exist outside the political world order, and the extreme injustice of the taking of innocent lives. The attacks were directed against two of the most famous buildings in America, with the Twin Towers seen as being representative of the West's affluence, success, and material bounty – all phenomena associated with Jupiter. The attacks were designed to make a big statement, to cause the maximum amount of disturbance, targeting buildings of enormous size or significance. In an event without precedent on the US mainland, the entire world was left to contemplate and try to comprehend what had happened, to struggle with a profound sense of injustice and outrage. The Jupiter archetype is also evident in the response to the attacks: the positive spirit of the

New Yorkers, the widespread messages of support and goodwill, the international response to the crisis. In these ways, the Eris archetype, as we have defined it, seems to have drawn out and activated the archetypal Jupiter, with the immense strife of these events calling forth, in the latter examples, the most positive, elevated aspects of the human spirit. Perhaps more than any other event, 9/11 brought to light that which needs to be confronted if the world is to realize its global unity, with the attacks prompting a flurry of international discussions amongst the world's nations, pulling the collective attention to the new reality of global terrorism, significantly contributing to closer international relations that will be essential to form an adequate planetary government for our time.

In the case of New Orleans, we find again that the 'wrath of nature' was inflicted upon one of America's most famous and most visited cities (relating to Jupiter). And again this tragedy called forth from many of those involved in the crises positive, elevated responses, again drew together large sections of the global community in solidarity with the victims.

So too the global financial crisis stirred in many people an acute sense of injustice given that the causes of the financial collapse were attributed to corporate greed, over-confidence, short-selling and profiteering on the stock market, and given the hundred-billion dollar bailouts of major banks by the taxpayer, which has left many national economies in dire straits. Reflecting other themes perhaps associated with the Eris-Jupiter combination, as outlined above, the global financial crisis can perhaps be seen as the resentment and retaliation against excess, abundance, the rich, and the affluent (all Jupiterian phenomena), and as a compensatory reaction to unchecked commerce and profiteering. It also marked, according to some commentators, the 'end of capitalism' – or at least a certain form of capitalism – and thus heralded a major turning point. And for all the injustice of this crisis, it has brought the world's economic powers into closer union than ever before, mobilizing a global response.

Again, certain themes in each of these crises can be understood in terms of other world transits taking place at the time, such as Saturn-Pluto in the case of 9/11, or Saturn-Neptune for the Indian Ocean tsunami and the New Orleans disaster. Indeed, it is imperative, I believe, that each of the above events be understood by taking into account these other world transits. It is also crucial to stress, of course, that not every natural disaster occurs in correlation with major transits of any of the planets, including Eris.

A further possible expression of Jupiter-Eris alignments might be seen in the bringing to justice of figures responsible for atrocities, or bringing atrocities to public awareness. For example, in the 1960s the growing resentment within America (and across Europe) against military involvement in Vietnam increased further after the brutal 1968 massacre at My Lai in which around 350 villagers were butchered by US troops (under a Mars-Saturn-Eris conjunction). Nothing was known of this atrocity for over one year after it took place, until it came to light between September and November of 1969 during the subsequent Jupiter-Eris conjunction. More recently, as noted above, the Kosovo genocide came to the attention of the international community in 1999 under a Jupiter-Eris conjunction, with Slobodan Milosovic indicted for crimes against humanity. Then, in July 2008, after years living a second life in disguise as a healer, Radovan Karadzic, former leader of the Bosnian Serbs, was finally captured and brought to justice under the Jupiter-Eris square, perhaps pointing to the association of the archetypal Eris with justice or the bringing to light of cases of injustice. More recently, Al-Qaeda leader Osama Bin Laden was finally tracked down and killed on May 1, 2011, under an exact transit of Jupiter to Eris (and also during a Saturn-Eris opposition). This was widely heralded as the attainment of some kind of justice for the 9/11 atrocity – indeed this was announced with the headline 'Justice is done'. It was also, of course, seen as a major turning point in the war on terror. Just a few weeks later, on May 26, 2011, under the same Jupiter-Eris conjunction, Serbian general Ratko Mladic, accused of the massacre of 7,500 Muslim men and boys at Srebrenica in 1995, was also arrested.

As with the Saturn-Eris examples, these possible correlations must be seen as entirely provisional and speculative. Extensive research will be required for any correlations to be corroborated.

Alignment	Dates	Events
Conjunction	March 1999–March 2001	Kosovo genocide Columbine shootings
Square	August 2001–July 2002	9/11 terrorist attacks
Opposition	July 2003–November 2005	Hurricane Katrina, New Orleans Indian Ocean tsunami Kashmir earthquake Terrorist bombings in London Bali bombings
Square	February 2008–January 2009	Mumbai terrorist massacres Global financial crisis Burma cyclone China earthquake
Conjunction	February 2011–February 2012	Japanese earthquake-tsunami-Fukushima disaster Greek financial crisis jeopardizes European Union Killing of Osama Bin Laden Capture of Ratko Mladić Norway twin terror attacks

Figure 19. Some Possible Correlations with Jupiter-ErisAlignments.
*15 degree orbs used for conjunctions and oppositions; 10 degree orbs used for squares.

APPENDIX D

Ephemeris Tables for Eris (1600–2036)

Ephemeris Table for Eris, 1600–1900
This Eris ephemeris is adapted from Astrodienst data available at www.astro.com, with the kind permission of Dr Alois Trendl.

One entry per decade

DATE (day, month, year)	LONGITUDE (degrees, sign, minutes)		
01.01.1600	27	Taurus	22
01.01.1610	3	Gemini	1
01.01.1620	9	Gemini	48
01.01.1630	18	Gemini	7
01.01.1640	28	Gemini	36
01.01.1650	13	Gemini	44
01.01.1660	28	Cancer	39
01.01.1670	18	Leo	6
01.01.1680	8	Virgo	16
01.01.1690	26	Virgo	48
01.01.1700	12	Libra	46
01.01.1710	26	Libra	32
01.01.1720	8	Scorpio	47
01.01.1730	20	Scorpio	12
01.01.1740	1	Sagittar	4
01.01.1750	11	Sagittar	41
01.01.1760	22	Sagittar	0
01.01.1760	22	Sagittar	0
01.01.1770	2	Capric	1
01.01.1780	11	Capric	33
01.01.1790	20	Capric	36
01.01.1800	28	Capric	57
01.01.1810	6	Aquar	38
01.01.1820	13	Aquar	39
01.01.1830	20	Aquar	2
01.01.1840	25	Aquar	48
01.01.1850	1	Pisces	5
01.01.1860	5	Pisces	53

01.01.1870	10	Pisces	16
01.01.1880	14	Pisces	20
01.01.1890	18	Pisces	4
01.01.1900	21	Pisces	35

Ephemeris Table for Eris, 1900–2036
January positions; one entry per year

DATE (day, month, year)	LONGITUDE (degrees, sign, minutes)		
01.01.1900	21	Pisces	35
06.01.1901	21	Pisces	56
01.01.1902	22	Pisces	15
06.01.1903	22	Pisces	36
01.01.1904	22	Pisces	55
05.01.1905	23	Pisces	16
10.01.1906	23	Pisces	37
05.01.1907	23	Pisces	55
10.01.1908	24	Pisces	15
04.01.1909	24	Pisces	33
09.01.1910	24	Pisces	53
04.01.1911	25	Pisces	11
03.01.1913	25	Pisces	48
08.01.1914	26	Pisces	8
03.01.1915	26	Pisces	26
08.01.1916	26	Pisces	46
02.01.1917	27	Pisces	3
07.01.1918	27	Pisces	22
02.01.1919	27	Pisces	39
07.01.1920	27	Pisces	58
01.01.1921	28	Pisces	15
06.01.1922	28	Pisces	34
01.01.1923	28	Pisces	50
06.01.1924	29	Pisces	8
10.01.1925	29	Pisces	27
05.01.1926	29	Pisces	43
10.01.1927	0	Aries	2
05.01.1928	0	Aries	18
09.01.1929	0	Aries	37
04.01.1930	0	Aries	53
09.01.1931	1	Aries	11
04.01.1932	1	Aries	27
08.01.1933	1	Aries	45
03.01.1934	2	Aries	1
08.01.1935	2	Aries	18
03.01.1936	2	Aries	34
07.01.1937	2	Aries	51
02.01.1938	3	Aries	7

07.01.1939	3	Aries	24
02.01.1940	3	Aries	40
06.01.1941	3	Aries	57
01.01.1942	4	Aries	13
06.01.1943	4	Aries	29
01.01.1944	4	Aries	45
05.01.1945	5	Aries	1
10.01.1946	5	Aries	18
05.01.1947	5	Aries	33
10.01.1948	5	Aries	50
04.01.1949	6	Aries	5
09.01.1950	6	Aries	21
04.01.1951	6	Aries	36
09.01.1952	6	Aries	53
03.01.1953	7	Aries	8
08.01.1954	7	Aries	24
03.01.1955	7	Aries	39
08.01.1956	7	Aries	55
02.01.1957	8	Aries	10
07.01.1958	8	Aries	25
02.01.1959	8	Aries	40
07.01.1960	8	Aries	55
01.01.1961	9	Aries	10
06.01.1962	9	Aries	25
01.01.1963	9	Aries	40
06.01.1964	9	Aries	56
10.01.1965	10	Aries	11
05.01.1966	10	Aries	26
10.01.1967	10	Aries	41
05.01.1968	10	Aries	56
09.01.1969	11	Aries	11
04.01.1970	11	Aries	25
09.01.1971	11	Aries	41
04.01.1972	11	Aries	55
08.01.1973	12	Aries	10
03.01.1974	12	Aries	24
08.01.1975	12	Aries	39
03.01.1976	12	Aries	54
07.01.1977	13	Aries	8
02.01.1978	13	Aries	23
07.01.1979	13	Aries	37
02.01.1980	13	Aries	51
06.01.1981	14	Aries	6
01.01.1982	14	Aries	20
06.01.1983	14	Aries	34
01.01.1984	14	Aries	48
15.01.1985	15	Aries	4
10.01.1986	15	Aries	17
05.01.1987	15	Aries	32

10.01.1988	15	Aries	46
04.01.1989	16	Aries	0
09.01.1990	16	Aries	15
04.01.1991	16	Aries	29
09.01.1992	16	Aries	43
03.01.1993	16	Aries	57
08.01.1994	17	Aries	11
03.01.1995	17	Aries	25
08.01.1996	17	Aries	39
02.01.1997	17	Aries	53
07.01.1998	18	Aries	7
02.01.1999	18	Aries	21
07.01.2000	18	Aries	35
01.01.2001	18	Aries	50
06.01.2002	19	Aries	3
01.01.2003	19	Aries	18
06.01.2004	19	Aries	31
10.01.2005	19	Aries	45
05.01.2006	19	Aries	59
10.01.2007	20	Aries	13
05.01.2008	20	Aries	27
09.01.2009	20	Aries	41
04.01.2010	20	Aries	55
09.01.2011	21	Aries	9
04.01.2012	21	Aries	23
08.01.2013	21	Aries	37
03.01.2014	21	Aries	51
08.01.2015	22	Aries	5
03.01.2016	22	Aries	19
07.01.2017	22	Aries	32
02.01.2018	22	Aries	46
07.01.2019	23	Aries	0
02.01.2020	23	Aries	14
06.01.2021	23	Aries	27
01.01.2022	23	Aries	42
06.01.2023	23	Aries	55
01.01.2024	24	Aries	10
05.01.2025	24	Aries	24
10.01.2026	24	Aries	37
05.01.2027	24	Aries	52
10.01.2028	25	Aries	5
04.01.2029	25	Aries	20
09.01.2030	25	Aries	33
04.01.2031	25	Aries	47
09.01.2032	26	Aries	1
03.01.2033	26	Aries	15
08.01.2034	26	Aries	29
03.01.2035	26	Aries	43
08.01.2036	26	Aries	57

Image Credits

Figure 1:
Wikipedia, 'Solar system'. http://en.wikipedia.org/wiki/File:Solar_System_ size_to_scale.svg. Adapted from original NASA image.

Figure 2:
Data derived from: http://solarsystem.nasa.gov/planets/profile. cfm?Object=Dwarf&Display=Moons; http://solarsystem.nasa.gov/planets/ profile.cfm?Object=KBOs&Display=OverviewLong; and http://www. chadtrujillo.com/quaoar/

Figure 4:
Data derived from http://solarsystem.nasa.gov/planets/index.cfm

Figure 5:
Courtesy NASA website. http://solarsystem.nasa.gov/multimedia/gallery/ Tilted_Eris.jpg

Figure 6:
The Judgment of Paris, Peter Paul Rubens, c. 1638–1639 (National Gallery, London)
source: http://commons.wikimedia.org/wiki/File:Peter_Paul_Rubens_115.jpg

Figure 7:
Eris (from inscription). Tondo of an Attic black-figure kylix (Altes Museum, Berlin)
Wikimedia Commons. 'Eris (mythology)'. http://commons.wikimedia.org/ wiki/File:Eris_Antikensammlung_Berlin_F1775.jpg.

Figure 15:
Jakob Böhme, *The Philosophical Globe* or *The Wonder-Eye of Eternity*
Mandala from 1647 English edition of Jakob Böhme's *XL Questions Concerning the Soule* (1620). Reproduced in C. G. Jung, T*he Archetypes and the Collective Unconscious*, Second Edition, p. 297.

Endnotes

Preface

1. The term *planet* has significantly different meanings in astronomy and astrology. In astrology, the meaning reflects the Greek origins of the term. As Richard Tarnas explains: 'The ancient Greek root for the word "planet" meant "wanderer" and signified not only Mercury, Venus, Mars, Jupiter, and Saturn but also the Sun and Moon, i.e., all the visible celestial bodies that, unlike the fixed stars, moved through the skies in ways that differed from the simple motion and eternal regularity of the diurnal westward movement of the entire heavens. Though a distinction is often made between planets and luminaries, the astrological tradition has generally retained the original more encompassing meaning, referring to the Sun and Moon as planets.' (Tarnas, *Cosmos and Psyche*, pp.504–5)

In astronomy, the term planet is defined more narrowly based on the specific physical attributes and orbital characteristics of planetary bodies, and must now be distinguished from new classes or subclasses of planetary objects called *dwarf planets* and *plutoids*. On August 26, 2006, the International Astronomical Union (IAU) made the following announcement:

'The IAU ... resolves that planets and other bodies in our Solar System be defined into three distinct categories in the following way:

(1) A planet is a celestial body that

(a) is in orbit around the Sun,

(b) has sufficient mass for its self-gravity to overcome rigid body forces so that it assumes a hydrostatic equilibrium (nearly round) shape, and

(c) has cleared the neighbourhood around its orbit.

(2) A "dwarf planet" is a celestial body that

(a) is in orbit around the Sun,

(b) has sufficient mass for its self-gravity to overcome rigid body forces so that it assumes a hydrostatic equilibrium (nearly round) shape,

(c) has not cleared the neighbourhood around its orbit, and

(d) is not a satellite.

(3) All other objects orbiting the Sun shall be referred to collectively as "Small Solar System Bodies".' (see http://www.iau.org/public_press/news/detail/iau0603/)

Consequently, the terms *planets* and *dwarf planets* were henceforth used to refer to two different classes of objects. Pluto, following criterion (c), was 'demoted' to the status of dwarf planet. The planets in the solar system are therefore now considered to be: Mercury, Venus, Earth, Mars, Jupiter, Saturn, Uranus, and Neptune.

On June 11, 2008, a further announcement was made by the IAU regarding the definition of a new sub-class of objects to be known as plutoids:

'Plutoids are celestial bodies in orbit around the Sun at a distance greater than that of Neptune that have sufficient mass for their self-gravity to overcome rigid body forces so that they assume a hydrostatic equilibrium (near-spherical) shape, and that have not cleared the neighbourhood around their orbit.' (See http://www.iau.org/public_press/news/detail/iau0804/)

Pluto and Eris are thus both considered to be dwarf planets and plutoids.

2. The fifth dwarf planet is Ceres, located in the asteroid belt between Mars and Jupiter. Because it is not a trans-Neptunian object, Ceres is not classified as a plutoid. However, a number of other recently discovered planet-like bodies are considered to be strong candidates for the plutoid classification, including Quaoar and Sedna.

3. According to the NASA website: 'Eris was first spotted in 2003 during a Palomar Observatory survey of the outer solar system by Mike Brown, a professor of planetary astronomy at the California Institute of Technology, Chad Trujillo of the Gemini Observatory, and David Rabinowitz of Yale University. The discovery was confirmed in January 2005, and was submitted as possible 10th planet in our solar system since it was the first object in the Kuiper Belt found to be bigger than Pluto.' In early 2011, however, the estimate of the size of Eris was revised. 'Eris first appeared to be larger than Pluto, a discovery that triggered debate in the scientific community and eventually led to the International Astronomical Union's decision in 2006 to clarify the definition of a planet. Recent observations indicate Eris may actually be a little smaller than Pluto.' Source: http://solarsystem.nasa.gov/planets/profile.cfm? Object=Dwa_Eris (accessed March 28, 2011). See Mike Brown's own account of the events surrounding Eris' discovery and Pluto's demotion in *How I Killed Pluto and Why It Had It Coming*.

4. See especially Richard Tarnas, *Cosmos and Psyche: Intimations of a New World View* for the empirical evidence in support of astrology. Further correlations are given in the Archai journal, edited by Le Grice and O'Neal. For more on the background to archetypal astrology and the wider academic discipline of archetypal cosmology, see Le Grice, 'The Birth of a New Discipline: Archetypal Cosmology in Historical Perspective', in *The Birth of a New Discipline, Archai: The Journal of Archetypal Cosmology*, Issue 1, 2009. (San Francisco: Archai Press, 2011).

5. The method employed to analyze and interpret the archetypal dynamics

of human experience in terms of the movements of the planets is based on a consideration of the geometric alignment – the specific angle of relationship – formed between the different planets in their respective orbits. The meaning of every planetary alignment or aspect depends both upon the archetypal characteristics associated with the planets involved and the particular angle of relationship between the planets. As in the Pythagorean view, in astrology principles of number and geometry are recognized as fundamental to the deep structure and organization of the cosmos, and these numeric principles are reflected in the geometric relationships between the planets.

The major aspects recognized in the astrological tradition are the conjunction (two or more planets approximately 0 degrees apart), the sextile (60 degrees), the square (90 degrees), the trine (120 degrees), and the opposition (180 degrees). Of these, the quadrature alignments – the conjunction, the opposition, and the square – are usually the most significant in terms of understanding both world events and the major themes of individual biography. In the astrological tradition, these alignments are considered to be dynamic, 'hard', or challenging in that they signify relationships between the archetypal principles that generally require some form of adaptation or considerable exertion or struggle to integrate, that tend to promote action to release the inherent energetic tension between the archetypal principles, and that are, therefore, often seen as most problematic or challenging, if ultimately creative and progressive. The trine and sextile, by contrast, are deemed 'soft', harmonious, or confluent aspects in that they tend to indicate a relatively well-established, already integrated, mutually supportive, and harmonious relationship between the archetypal principles. At the risk of oversimplification, one can think of the soft aspects as already integrated states of being and the hard aspects as dynamic states of becoming that require integration.

6. For more detail on Tarnas' understanding of the nature of archetypes, see *Cosmos and Psyche*, pp.80–101.

7. For more detail on the synthesis of Jungian psychology and the new paradigm sciences, see Le Grice, *The Archetypal Cosmos: Rediscovering the Gods in Myth, Science and Astrology* (Edinburgh: Floris Books, 2010).

Introduction

1. The discovery of the planetoid Chiron in 1977 provides another example, albeit on a lesser scale of significance, of this process. See also, notes 1 and 2 to Chapter 6.

2. Jung, *Psychological Types*, p.476.

3. According to Morin, 'The Planetary Era begins with the discovery that the Earth is a planet and with the entering into communication among various parts of the planet.' This occurred between 'the conquest of the Americas and the Copernican Revolution.' The first globe appeared, Morin reports, in 1492 (*Homeland Earth: Manifesto for a New Millennium*, p.6). See also, Sean Kelly, *Coming Home: The Birth and Transformation of the Planetary Era*, pp. vii–viii.

Chapter 1

1. Dane Rudhyar, *Astrology of Personality*, pp.241–242 (Rudhyar's emphasis).

2. Jung, *On the Nature of the Psyche*, p.145.

3. See Le Grice, *Archetypal Cosmos*, chapter 6, 'The archetypal order', pp.152–178, for more detail on Jung's understanding of archetypes in relation to astrology.

4. See Le Grice, *Archetypal Cosmos*, pp.196–205.

5. Pluto is connected with instinctual compulsion, Neptune with the loss of individual separateness and preconscious fusion with the environment, and Uranus with unrestrained instinctual freedom. The archetypal principles associated with the outer planets play an important part in all human experience, but to engage consciously with the highest potentials of these archetypes is extremely challenging. Pluto, for example, is associated with the sexual drives and with the cycle of birth-sex-death that are obviously fundamental to all human experience. Yet it is also associated with the process of psychospiritual transformation, relating to spiritual rebirth and to the purification or purgation process that is fundamental to the life of the mystic.

6. Although Eris is currently extremely distant from the Sun, its tilted elliptical orbit (46.87 degrees from the ecliptic) means that in less than three hundred years from now it will move as close to the Sun as Neptune. That Eris comes within the orbits of Pluto and Neptune at certain points in its own orbit further reinforces the sense that the archetypes associated with all three planets are closely connected in meaning, each relating, I believe, to the underlying ground of reality and the deepest dimensions of psychospiritual transformation. For details on the orbital inclination of Eris, see http://solarsystem.nasa.gov/planets/profile.cfm?Object=Dwa_Eri s&Display=Facts&System=Metric.

Chapter 2

1. This phenomenon has been discussed by a number of astrologers and scholars, including Dane Rudhyar, Liz Greene, Gerry Goddard, and Richard Tarnas.

2. According to Jung, 'Synchronicity therefore means the simultaneous occurrence of a certain psychic state with one or more external events which appear as meaningful parallels to the momentary subjective state' (Jung, *Synchronicity: An Acausal Connecting Principle*, p.36).

3. The 1848 revolutions took place during the Uranus-Pluto conjunction of 1845, an alignment often coinciding with revolutionary activity, rebellion, and social tumult and turbulence. See Tarnas, *Cosmos and Psyche*, pp.141-193.

4. See, for example, Tarnas, *Cosmos and Psyche*, pp.92–100.

5. Admittedly, this task is made considerably more difficult because several new planet-like bodies have been discovered in recent years, and one cannot therefore readily identify which events relate to which of the new planets. Against this background, I take all the major events of recent years to be potentially relevant for determining Eris' meaning and use the other approaches (set forth in the other chapters of this book) to more accurately gauge which particular themes might relate specifically to Eris. As I see it, each of the new planetary bodies might be archetypally connected to major recent events, but each to different themes and motifs within those events.

6. I am grateful to Sean Kelly for raising the possible connection between Eris, globalization, and the Planetary Era.

7. Swimme, 'Cosmic creation story,' in *Readings in Ecology and Feminist Theology*, p.250.

8. We might note here the possible symbolic connection between the apple of the biblical Fall in the Garden of Eden and, in Greek mythology, the apple rolled by the goddess Eris at the feet of three goddesses during the wedding party. See Chapter 4, pages 67–69. In Romantic literature, the mythic fall that leads to the separation of ego-consciousness from nature is seen as fundamental to psychospiritual evolution. According to M.H. Abrams: 'the fall from primal unity into self-division, self-contradiction, and self-conflict ... [is] regarded as an indispensable first step along the way toward a higher unity which will justify the sufferings undergone en route. The dynamic of the process is the tension toward closure of the divisions, contraries, or "contradictions" themselves' (*Natural Supernaturalism*, p. 255). This process, as we will see later, appears to be connected to themes associated with the Eris archetype.

9. Jung, *Undiscovered Self*, p.111.

10. Schopenhauer, *World as Will and Idea*, pp.73–74.

Chapter 3

1. Frank Poletti gives further detail on the process by which astrologers determined the set of meanings associated with Neptune in his forthcoming doctoral dissertation, 'Neptune and the Nineteenth Century' (California Institute of Integral Studies, San Francisco).

2. Dane Rudhyar, however, anticipated the discovery of at least one other planet beyond Pluto. Isabel Hickey also speculated that Pluto was part of a pair of planets, and the that the corresponding planetary archetype therefore had a dual nature. See Rudhyar's attributions to Pluto in *The Sun is Also a Star: The Galactic Dimensions of Astrology*; and Hickey's description of the dual nature of Pluto in *Astrology: The Cosmic Science*.

3. Freud, *Civilization and its Discontents*, p.42.

4. Teilhard de Chardin, *Phenomenon of Man*, p.240.

5. See especially Erich Neumann, *Origins and History of Consciousness*.

6. As Neumann notes, 'Through the heroic act of world creation and division of opposites, the ego steps forth from the magic circle of the uroborus [primal unity] and finds itself in a state of loneliness and discord' (*Origins and History of Consciousness*, p.114).

7. Something like this pattern is also evident in Sri Aurobindo's understanding of the course of human spiritual evolution in which the *nirvana* experience (relating to Uranus [liberation] and Neptune [spirituality]) is not seen as the end point of spiritual evolution but is to be followed by the transformation (relating to Pluto) of the mental, vital, and material dimensions of being. See Aurobindo, *Human Cycle: The Ideal of Human Unity* and *Integral Yoga*.

8. Jung, *Aion*, p.71.

Chapter 4

1. For a description of the Ouranos myth, see Graves, *Greek Myths*, section 6a, vol 1., p.37. For Tarnas' discussion of this topic, see *Prometheus the Awakener*, pp.10–16. In *Cosmos and Psyche*, his later work, Tarnas identifies certain parallels between the character of the astrological Uranus and the mythic figure of the same name. Both are associated, he notes, with 'the cosmic and celestial, with space and space travel, and with astronomy and astrology' – these themes reflecting the mythic Ouranos' role as god of the 'starry sky' (*Cosmos and Psyche*, p.94).

2. Graves, *Greek Myths*, section 11a, vol. 1, p.49 and p.36.

3. As Erich Neumann puts it: 'The coming of consciousness ... is the real "object" of creation mythology' (*Origins and History of Consciousness*, p.16).

4. Freud, *Civilization and its Discontents*, p.49.

5. In his article 'Eris Stirs up Trouble' in *The Mountain Astrologer* (April/May 2007), Zane Stein connects Eris to the myth of Persephone, pointing out that Mike Brown had hoped to use this name for the new planet until he discovered it had already been used for another astronomical object.

6. Graves, *Greek Myths I*, section 19a, p.73.

7. Graves, *Greek Myths I,* section 81n, p.271.

8. See Graves, *Greek Myths II,,* section 159, pp.269–275.

9. Graves, *Greek Myths II*, section 159e, p.269.

10. The feud also relates to the ownership of the Golden Fleece. Considering this aspect of the story, I am reminded of the Iroquois myth of world creation arising out of the dispute between two brothers. Perhaps this creation myth also points to the inherent discord we have associated with the Eris archetype as the root archetypal factor behind human civilization.

11. Graves, *Greek Myths II*, section 111e, p.45.

12. Washburn, 'The pre/trans fallacy reconsidered', in *Ken Wilber in Dialogue*, pp.70–71.

13. See Jung, *Undiscovered Self*, p.110. For a discussion of this topic in relation to archetypal astrology, see Le Grice, *Archetypal Cosmos*, pp.39, 263–300.

14. Jung, *Mysterium Coniunctionis*, p.471.

15. Jung, *Undiscovered Self*, p.112.

16. Martin, *Meaning of the 21st Century*, Part One, pp.5–6 , and Part Three.

17. Graves, *Greek Myths II*, p.269.

18. Harding, *Animate Earth*, p.209.

19. See Hesiod *Works and Days*, p.90ff; and *Theogony*, p.226ff.

20. Because human beings habitually emphasize good over evil, pleasure over pain, happiness over suffering, there arises a one-sidedness in human experience that typically excludes the negative half of experience. If Eris is related to a principle of counterbalancing reaction, then in its activity to maintain harmony and balance Eris might well be often expressed, at least initially, in its negative pole, as the counterbalancing expression of the darker, more painful aspects of experience.

21. Jung, *Two Essays on Analytical Psychology*, pp.188–211.

22. It seems possible to me that this dream also symbolized the astronomical demotion of Pluto.

23. In Greek myth, Eris is the daughter of dark night (*Nyx*).

24. One might question, of course, whether it is possible that anything can be more powerful or fundamental than the instinctual-evolutionary

force associated with the Pluto archetype. In psychological terms, however, it is important to keep in mind that instinctual power increases in direct coincidence with the psychological division accompanying the emergence of ego-consciousness – a division that I am tentatively connecting here to the Eris archetype. The accumulation of instinctual power in the psyche appears to be a function of our loss of wholeness, of the repression of the instinctual power of the ground of being that permits ego-consciousness to emerge. (See Washburn, *The Ego and Dynamic Ground* for one account of this perspective.) The power of the id, the instinctual unconscious, increases the more it is repressed. Conversely, then, through the progressive recovery of wholeness at a higher level, instinctual power dissipates as it is swallowed up in a greater totality. It is in this sense that Eris is more fundamental than Pluto; the power of the instincts associated with Pluto are dependent on the expression of the Eris archetype.

Chapter 5

1. For Empedocles, love and strife are the two fundamental 'soul substances'. See *Greek Philosophy: Thales to Aristotle*, pp.49–50.
2. Cornford, *From Religion to Philosophy*, p.66.
3. Ibid.
4. Ibid., p.70.
5. Ibid., p.64.
6. Ibid.
7. Ibid., pp.64–65.
8. Ibid., p.63.
9. Ibid., p.190.
10. Ibid., p.191.
11. Ibid., p.190.
12. Saotome, *Aikido and the Harmony of Nature*, p.69.
13. Ibid., p.68.
14. Ibid.
15. More specifically, it seems possible that Eris might have some connection to adaptive responses to new circumstances and sudden revelations of major evolutionary turning points, especially when it is in major alignment to Uranus. Teilhard de Chardin, one of the primary theorists of evolution, was born with Uranus in opposition to Eris and Mars, and the Moon square to Eris and Mars. Given our hypothesis, it seems entirely fitting that Teilhard was inspired by his experiences in World War II (relating to Mars and its association with the warrior archetype) to intuit the possibility of the realization of a fuller mode of human existence characterized by

a greater 'zest' for life, and to develop his sense of greater human and planetary unity. It is perhaps not insignificant either that Charles Darwin formulated and published his theory of evolution during a long personal transit of Eris, first to his Sun and then to Mercury. See Mark Williams insightful essay on the possible meanings of Eris at http://mvtabilitie. blogspot.com/2009/04/astrological-eris.html.   Williams   reports:

> Transiting Eris conjuncted Darwin's Sun in Aquarius as he set off on his voyage aboard HMS Beagle, during which he made the observations that led eventually to the writing of *The Origin of Species*. The conjunction was operative – to within a degree – during the entire 1831–6 voyage. As *The Origin of Species* was published in 1859, Eris moved in to conjoin Darwin's natal Mercury and precisely square his natal Saturn-Neptune conjunction in Sagittarius in the 12th [house]. The book's publication had an interesting Eridian flavour: Alfred Russel Wallace sent Darwin a theoretical version of a mechanism for evolution, closely similar to his own, just as Darwin was preparing to publish. Fortunately the two were able to come to an agreement to present their work together at a meeting of the Linnean Society of London in July 1858.

It is worth pointing out here, too, that Jupiter was exactly conjunct Eris in December 1831 when Darwin set sail, perhaps suggesting the long journey to foreign lands (Jupiter) that was to initiate a major turning point (Eris) in our understanding of the origins and evolution of human life. Darwin's discovery of evolution is also connected to the Uranus archetype. See Tarnas, *Prometheus the Awakener*, pp.37–38.

16. Jung was born with an exact trine aspect between Mercury and Eris, and the insight into the relationship between opposites informs all his psychological theories – as in his psychological typology, his study of the opposites in alchemy, his admiration for Heraclitus' ideas, his theory of the compensatory function of dreams within the psyche as it strives to maintain dynamic harmony, the dialectic between ego and unconscious during individuation, and the one-sidedness of intellectual development compensated by a reaction from the unconscious – which, if our thesis is correct, are all consistent with the meaning of a Mercury-Eris alignment. It is perhaps also relevant that it was his intellectual disagreement with Freud over the aetiology of neuroses (relating to Mercury and its association with the intellect and ideas) that was the primary cause of the major dispute and turning point in Jung's life.

17. *Gospel of Thomas*, Logion 16, p.19.

18. Nietzsche, 'Genealogy of Morals', section 5, in *Basic Writings of Friedrich Nietzsche*, p.467.

19. Nietzsche, 'Beyond good and evil', in *Basic Writings of Friedrich Nietzsche*, p.193.

20. Nietzsche, *Basic Writings*, p.125. Nietzsche also had Eris in major alignment to the Sun and Venus in his chart, which we can perhaps see in his word choice in proclamations such as: 'You must be proud of your enemy: then the success of your enemy shall be your success too' and 'let us also be enemies, my friends! Let us divinely strive against one another!' Breaking these expressions down astrologically, the Sun relates to pride, Venus to friendships, and Neptune to the divine. And Eris, in light of our thesis, might well relate to moving beyond the opposites of enmity and friendship to higher levels of realization, and to using competition as a spur to excellence – thus divinely striving, taking pride in enmity and opposition in the face of the other, and holding that approach up as an ideal of friendship.

21. Hesiod, *Works and Days*, p.1. Commenting on this passage, Mark Williams suggests that the astrological Eris might be connected to the emotion of envy:

> However, it's striking that Hesiod's attempt to demarcate the twin Erides as having 'wholly different natures' wobbles somewhat towards the end of his description in the *Works and Days*. The vision of the angry potters and resentment-filled beggars is taking the idea of rivalry (positive Eris) and moving it towards quarrelling and violence (negative Eris.) The thing they seem to share, the archetypal core, if you like, is *envy*. As Liz Greene has pointed out, in a discussion of Saturnian defences, '[e]nvy can be extremely creative. Through making envy conscious, we can discover what we want and value, because we see it in someone else and wish we had it . . . . Envy, recognized and constructively channelled, can spur us toward developing qualities and abilities which we might not otherwise have recognized as our own potentials.' (*Barriers and Boundaries: The Horoscope and the Defences of the Personality*, p.139.) Envy, if we cannot acknowledge it, can also lead us to attempt to destroy the person or institution which possesses the quality we feel we do not have ourselves. The positive Eris is actually called 'Emulation' elsewhere in classical literature, and I think the concepts of conscious and unconscious envy may well be the key to the astrological Eris. (http://mvtabilitie.blogspot.com/2009/04/astrological-eris.html)

To this I would simply add that the experience of injustice or perceived lack of fairness is often the basis for the envy and resentment Williams mentions here – the sense that it is unfair that another person has something that you yourself feel you deserve.

22. Aurobindo, *Future Evolution of Man*. Sri Aurobindo was born with

Mercury (relating to the mind, perception, thought, and communication) in opposition to Eris.

23. Watts, *Two Hands*, pp.42–43.

24. Ibid., p.44.

25. Ibid., pp.49–71.

26. Ibid., p.46.

27. Richard Tarnas has made a number of these observations with regard to the expression of the Saturn-Pluto archetypal complex in the life of Thomas Berry.

28. Berry, *Dream of the Earth*, p.217.

29. Ibid., pp.218–219.

30. Ibid., p.219 (Berry's emphasis).

31. Ibid. (Berry's emphasis)

32. Ibid.

33. Ibid., p.220.

34. Ibid., p.221.

35. Ibid., p.223.

36. Abrams, *Natural Supernaturalism*, pp.257–58.

37. Ibid., p.258.

38. Ibid., p.259.

39. Ibid., p.260.

40. Blake, *Complete Poetry and Prose of William Blake*, p.34.

41. Abrams, *Natural Supernaturalism*, p.260.

42. Swimme and Berry, *Universe Story*, p.215.

43. See Appendix B for details of correspondences between these events and astrological world transits – the study of correlations between planetary alignments and archetypal themes evident in world history over time.

44. See Appendix C on possible correlations between Jupiter-Eris transits and certain prominent natural disasters and global terrorist attacks.

45. Tarnas, *Cosmos and Psyche*, pp.470–478. See also, Tarnas, 'World transits 2000–2020: an overview', pp.146–51. http://www.archaijournal. org/Tarnas_World_Transits_2000-2020_An_Overview.pdf

46. Tarnas, *Cosmos and Psyche*, pp.209–288.

47. Tarnas, 'World transits 2000–2020: an overview', p.165. http://www. archaijournal.org/Tarnas_World_Transits_2000-2020_An_Overview.pdf

48. See BBC article, 'Pandora galaxy structure crash yields dark matter clues', June 22, 2011, http://www.bbc.co.uk/news/science-environment-13878171.

49. Ibid.

50. 'Science in the underworld', BBC website article, April 23, 2003. http:// news.bbc.co.uk/1/hi/sci/tech/2981837.stm (accessed March 25, 2011).

Chapter 6

1. It might well be that newly discovered bodies might represent more specific archetypal functions and principles that do not have exclusive rulership of any specific signs or houses or that relate only to certain characteristics and qualities of particular signs. The other trans-Neptunian dwarf planets (plutoids) might, like Eris, be in some way connected to the generative source, the creative ground of reality, and to the original unity before the diasphora of the human species across the globe and the coming into existence of separate races, nations, cultures, and civilizations. Research into possible mythic parallels with the names of new planetary bodies is somewhat more problematic now because astronomers have deviated from the tradition of naming new planetary bodies after Greek deities, drawing instead on more obscure figures from other mythic traditions. Sedna, for instance, is the Inuit goddess of the sea; Makemake is a fertility god and creator of humanity in the Rapa Nui mythology of the peoples of Easter Island; and Haumea is an Hawaiian goddess of fertility and childbirth. The diversity of origin in itself might be symbolically significant, of course, indicative of the global culture in which we now live, and reflecting the transcultural universality of the archetypal principles studied in astrology.

2. For more information on Chiron, see Melanie Reinhart, *Chiron and the Healing Journey* and Barbara Hand Clow, *Chiron: Rainbow Bridge Between the Inner and Outer Planets.*

3. Greene, *Astrology of Fate*, p.222.

4. Ibid.

5. Ibid., p.226.

6. Ibid.

7. Ibid.

8. The French name for the sign Libra is *La Balance.*

9. Jung, *Archetypes of the Collective Unconscious*, p.381, par. 704.

10. Ibid., p.341.

11. Ibid., p.381.

12. See Wikipedia, 'Discordianism', http://en.wikipedia.org/wiki/Discordianism#Original_Snub

13. Macalypse the Younger, *Principia Discordia*, p.21.

14. See http://www.petitiononline.com/ffhoeris/

15. See 'Proof by Contradiction, Wikipedia entry, http://en.wikipedia.org/wiki/Proof_by_contradiction

Conclusion

1. Ross, *Gospel of Thomas*, logion 50, p.38.

2. Jung, *Memories, Dreams, Reflections,* p.340.

Bibliography

Abrams, M.H. (1973) *Natural Supernaturalism: Tradition and Revolution in Romantic Literature*. New York: W.W. Norton & Company.

Allen, Reginald. E. ed. (1991) *Greek Philosophy: Thales to Aristotle*. Third Edition. New York: Free Press.

Aurobindo Ghose, Sri (1962) *The Human Cycle; The Ideal of Human Unity; War and Self-Determination*. Pondicherry, Sri Aurobindo International Centre of Education.

—, (1962) *The Future Evolution of Man*. Prepared by E.B. Saint-Hilare. http://www.mountainman.com.au/auro_1.html (accessed 03/08/2011).

—, (1993) *The Integral Yoga: Sri Aurobindo's Teaching and Method of Practice. Selected Letters of Sri Aurobindo*. Pondicherry, India: Sri Aurobindo Ashram Press.

Becker, Ernest (1973) *The Denial of Death*. New York: The Free Press.

Berry, Thomas (1990) *The Dream of the Earth*. San Francisco: Sierra Club Books.

Blake, William (1988) *The Complete Poetry and Prose of William Blake*. Newly Revised Edition. Edited by David V. Erdman. New York: Anchor Books.

Brown, Mike (2010) *How I Killed Pluto and Why It Had It Coming*. NewYork: Spiegel & Grau.

Clow, Barbara Hand (1987) *Chiron: Rainbow Bridge Between the Inner and Outer Planets*. St. Paul, MN: Llewellyn Publishers.

Cornford, F.M. (1957) *From Religion to Philosophy: A Study in the Origins of Western Speculation*. New York: Harper & Row.

Freud, Sigmund. (1930/1989) *Civilization and its Discontents*. The Standard Edition. Translated by James Strachey. Repr. New York: W.W. Norton & Co.

Graves, Robert (1990) *The Greek Myths I & II*. Revised Edition. Repr. London: Penguin.

Greene, Liz (1997) *The Astrology of Fate: Fate, Freedom and Your Horoscope*. 1984. Repr. London: Thorsons.

Harding, Stephan (2009) *Animate Earth: Science, Intuition and Gaia*. Second Edition. Totnes, Devon, UK: Green Books.

Hesiod (2007) *Works and Days*. Forgotten Books. http://books.google.com/books?id=MGLUjHezCRYC&dq= hesiod+works+and+days&q=%28c%29#v=snippet&q=(c)&f=false (accessed 25/03/2011).

Hickey, Isabel M. (1992) *Astrology: A Cosmic Science*. Sebastopol, CA: CRCS.

Jung, Carl Gustav (1953–1979) *The Collected Works of C. G Jung*. 19 vols. Trans. R.F.C. Hull Princeton: Princeton University Press and London: Routledge & Kegan Paul.

—, (1921/1971). *Psychological Types*. A Revision by R F.C. Hull of the Translation by H.C. Baynes. Vol. 6 of *The Collected Works of C.G. Jung*. Trans. R.F.C. Hull. Repr. London: Routledge.

—, (1954) 'On the nature of the psyche.' In *On the Nature of the Psyche*. Trans. R.F.C. Hull, pp. 81–171. Repr. Abingdon, UK: Routledge.

—, (1955/1977) *Synchronicity: An Acausal Connecting Principle*. Trans. R.F.C. Hull. Repr. London: Routledge.

—, (1955–1956/1989) *Mysterium Coniunctionis*. 2nd ed. Vol. 14 of *The Collected Works of C.G. Jung*. Trans. R.F.C. Hull. Repr. Princeton: Princeton University Press.

—, (1958/2006) *The Undiscovered Self.* Trans. R.F.C. Hull. Repr. New York: Signet.

—, (1963/1983) *Memories, Dreams, Reflections*. Edited by Aniele Jaffe. Trans. Richard Wilson and Clara Wilson. Repr. London: Flamingo.

—, (1966/1990). *Two Essays on Analytical Psychology*. 2nd ed. Vol. 7 of *The Collected Works of C.G. Jung*. Trans. R.F.C. Hull. Repr. London: Routledge.

—, (1968) *The Archetypes and the Collective Unconscious*. 2nd ed. Vol. 9, part I of *The Collected Works of C.G. Jung*. Trans. R.F.C. Hull. Princeton: Princeton University Press.

—, (1968). *Aion: Researches into the Phenomenology of the Self.* 2nd ed. Vol. 9, part II of *The Collected Works of C.G. Jung*. Trans. R.F.C. Hull. London: Routledge.

—, (1970) *Civilization in Transition*. 2nd ed. Vol. 10 of *The Collected Works of C.G. Jung*. Trans. R.F.C. Hull. Princeton: Princeton University Press.

—, (2004) *On the Nature of the Psyche*. Trans. R.F.C. Hull. Repr. Abingdon, UK: Routledge.

Kelly, Sean (2010) *Coming Home: The Birth and Transformation of the Planetary Era*. Great Barrington, MA: Lindisfarne Books.

Kelly, Sean, & Donald Rothberg, eds. (1998) *Ken Wilber in Dialogue: Conversations with Leading Transpersonal Thinkers*. Wheaton, IL: Quest Books.

Le Grice, Keiron (2010) *The Archetypal Cosmos: Rediscovering the Gods in Myth, Science and Astrology*. Edinburgh: Floris Books.

—, (2011) 'The birth of a new discipline: archetypal cosmology in

historical perspective'. In *The Birth of a New Discipline: An
Introduction to Archetypal Cosmology, Archai: The Journal of Archetypal
Cosmology*, issue 1, summer 2009. Edited by Keiron Le Grice and Rod
O'Neal. San Francisco: Archai Press.

Macalypse the Younger (1980) *Principia Discordia Or, How I Found
Goddess and What I Did to Her When I Found Her: The Magnum
Opiate of Macalypse the Younger.* Loompanics Unlimited.

Martin, James (2006) *The Meaning of the 21st Century: A Vital Blueprint
for Ensuring our Future.* New York: Riverhead Books.

Morin, Edgar (1999) *Homeland Earth: A Manifesto for the New
Millennium.* Translated by Sean M. Kelly and Roger Lapointe.
Cresskill, NJ: Hampton Press.

Neumann, Erich (1954/1973) *The Origins and History of Consciousness.*
Repr. Princeton: Princeton University Press.

Nietzsche, Friedrich (1968) *Thus Spoke Zarathustra.* Translated by
Richard J. Hollingdale. New York: Penguin.

—, (2000) *Basic Writings of Nietzsche.* Translated by Walter Kaufmann.
New York: The Modern Library.

Parker, Philip (2010) *World History.* London: Dorling Kindersley.

Reinhart, Melanie (2010) *Chiron and the Healing Journey: an Astrological
and Psychological Perspective.* Third edition. Starwalker Press.

Ross, Hugh McGregor, trans. (2002) *The Gospel of Thomas.* London:
Watkins Publishing.

Rudhyar, Dane (2004) *The Galactic Dimension of Astrology: The Sun is
Also a Star.* Santa Fe, New Mexico: Aurora Press.

Saotome, Mitsugi (1993) *Aikido and the Harmony of Nature.* Boston,
MA: Shambhala Publications.

Schopenhauer, Arthur (1995) *The World as Will and Idea.* Abridged in
one volume. London: Everyman.

Swimme, Brian (1995) 'The cosmic creation story.' In *Readings in Ecology
and Feminist Theology.* Edited by M.H. MacKinnon and M. McIntyre.
Kansas City, MO: Sheed & Ward.

Swimme, Brian, & Thomas Berry (1994) *The Universe Story: From
the Primordial Flaring Forth to the Ecozoic Era.* Repr. San Francisco:
HarperSanFrancisco.

Tarnas, Richard (1995) *Prometheus the Awakener: An Essay on the
Archetypal Meaning of the Planet Uranus.* Repr. Woodstock, CT:
Spring Publications, Inc.

—, (2006) *Cosmos and Psyche: Intimations of a New World View.* New
York: Viking.

Teilhard de Chardin, Pierre (1965) *The Phenomenon of Man.* New York:
Harper & Row.

Washburn, Michael (1995) *The Ego and the Dynamic Ground: A Transpersonal Theory of Human Development*. 2nd Edition. Albany: State University of New York Press.

—, (1998) 'The pre/trans fallacy reconsidered'. In *Ken Wilber in Dialogue: Conversations with Leading Transpersonal Thinkers*. Edited by Sean Kelly and Donald Rothberg. Wheaton, IL: Quest Books.

Watts, Alan (1963/1978) *The Two Hands of God: The Myths of Polarity*. London: Rider & Company.

Williams, Mark (2009) 'Astrology of Eris'. http://mvtabilitie.blogspot. com/2009/04/astrological-eris.html (accessed 28/03/2011).

Index

Praise for *The Archetyal Cosmos*

'Le Grice has a gift, perhaps even a genius, for extremely clear assessments, expositions, and formulations of complex ideas – all grounded in a deeper vision of things, which makes this clarity possible.'
RICHARD TARNAS, AUTHOR OF *THE PASSION OF THE WESTERN MIND AND COSMOS AND PSYCHE*

'Keiron Le Grice's book is a fantastic achievement, combining complex theories from leading figures in psychology and the physical sciences – especially Carl Jung, Pierre Teilhard de Chardin, and David Bohm. Le Grice's original synthesis demonstrates how profoundly these figures belong together in a new unified world view. With both the excellence of the writing and the high-powered nature of the ideas, *The Archetypal Cosmos* is destined to be an essential element in the contemporary planetary canon.'
BRIAN SWIMME, PROFESSOR OF COSMOLOGY, CALIFORNIA INSTITUTE OF INTEGRAL STUDIES

'Humanity has much to gain from this synthesis between planetary cycles and the dynamics and patterns of human experience.'
RUTH PARNELL, *NEXUS*

'The book is a brilliant synthesis of new ideas.'
DAVID LORIMER, *SCIENTIFIC MEDICAL NETWORK REVIEW*,

'Keiron Le Grice's *The Archetypal Cosmos* has an immense amount to offer, having given this reviewer much food for thought.'
JACK HERBERT, TEMENOS ACADEMY

The Archetyal Cosmos

Keiron Le Grice

 Also available
as an eBook

The modern world is passing through a period of critical change on many levels: cultural, political, ecological and spiritual. We are witnessing the decline and dissolution of the old order, the tumult and uncertainty of a new birth. Against this background, Keiron Le Grice argues that the developing insights of a new cosmology could provide a coherent framework of meaning to lead us beyond the growing fragmentation of culture, belief and personal identity.

In a compelling synthesis of the ideas of seminal thinkers from depth psychology and new paradigm science, Le Grice positions the new discipline of archetypal astrology at the centre of an emerging world view that reunifies psyche and cosmos, spirituality and science, mythology and metaphysics, enabling us to see mythic gods, heroes and themes in a fresh light.

Heralding a 'rediscovery of the gods' and the passage into a new spiritual era, *The Archetypal Cosmos* presents a new understanding of the role of myth and archetypal principles in our lives, one that could give a cosmic perspective and deeper meaning to our personal experience.

www.florisbooks.co.uk